Fourier Series and Transforms

A Computer Illustrated Text

FOURIER SERIES AND TRANSFORMS

R D Harding

University of Cambridge

Published in 1985 by
Taylor & Francis Group
270 Madison Avenue
New York, NY 10016

Published in Great Britain by
Taylor & Francis Group
2 Park Square
Milton Park, Abingdon
Oxon OX14 4RN

International Standard Book Number-10: 0-8527-4809-4 (Softcover)
International Standard Book Number-13: 978-0-8527-4809-1 (Softcover)
Consultant Editor: **Professor R F Streater** University of London

Library of Congress Cataloging-in-Publication Data

Catalog record is available from the Library of Congress

Taylor & Francis Group
is the Academic Division of Informa plc.

Visit the Taylor & Francis Web site at
http://www.taylorandfrancis.com

⟩ Contents

⟩ Preface

There are several topics in mathematics at advanced secondary school level or early university course level whose teaching I have always thought would benefit by being more extensively illustrated than is possible in conventional texts. Many such topics are of interest not only to specialist mathematicians but also to scientists and engineers, and from many years experience of teaching them, what seems to be lacking from the normal treatment is a visual and qualitative feel for the properties of the mathematics.

The widespread and rapid growth of cheap microcomputers means that it is possible to try to remedy this lack through the use of computer graphics. In this *computer illustrated text* normal figures and diagrams are replaced with screen displays which can be produced dynamically by the reader, offering a far greater variety of illustrations than would be possible with printed diagrams alone. Of course I have had to assume that the reader has access to, and is reasonably familiar with one of the computers and its programming language which I have chosen to write for.

In principle the idea could be applied to an entire textbook but it is felt that to start with it is better to concentrate on a few topics for which this treatment seems especially useful, and also to concentrate on imparting an overall feel for these topics since it is not being suggested that existing texts are inadequate in technical detail. This text is therefore meant to be complementary to a standard textbook or lecture course.

Some may wonder whether it would not be better to have the computer reproduce the text as well as the diagrams. There certainly are such packages available, but unless the topic is either very elementary or capable of being discussed very briefly indeed, the present state of computer technology does not allow these to run on a cheap home microcomputer such as the BBC Microcomputer or the Apple II. Another problem is that mathematics requires a wide range of special symbols and again the technology is not widely and cheaply available which would allow these symbols to be shown on a screen. There is also the fact that a printed text can be more easily browsed through.

These considerations have led to the design of this monograph, which it is hoped will be the first of several such titles. As it is experimental, comments and criticisms will be particularly welcome. I hope it will make learning more interesting and enjoyable for you, the reader. Good luck.

RDH
March 1984

〉 Acknowledgments

The author gratefully acknowledges the hospitality and help of the Algebra Institute, University of California, Santa Barbara, where much of this text was written.

The author also acknowledges Acornsoft Ltd for permission to use routines from *Graphs and Charts on the BBC Microcomputer*.

$\rangle$ Introduction

Fourier series and Fourier transforms form the basis of many powerful mathematical methods, useful in every quantitative science. Yet, because they are usually taught from a purely analytical standpoint, they are often found to be a difficult topic and are poorly understood. Even those who can carry out the analytical manipulations required in this subject often have problems obtaining a qualitative understanding.

This book is intended to help in giving a qualitative feel for the properties of Fourier series and Fourier transforms by using the illustrative powers of computer graphics. The computer programs are not intended to be complete, self-contained packages; they should be thought of as an integral part of the text and the student will derive considerable benefit from studying the structure of the programs themselves. It is intended that the text be read whilst sitting at the computer, so that examples can be run when suggested in the text.

The book is not intended to be a replacement for other forms of instruction, and in particular will not teach the analytical skills required to calculate Fourier coefficients or transforms. However, the insight obtained from the computations should prove a valuable aid to developing these skills; in particular it should encourage students to consider the qualitative aspects of their results, such as symmetry, asymptotic behaviour, etc, which so often provide a useful check on the validity of analytic results.

The programs have been designed to be 'open' to modifications. In the text, users are encouraged to read the source programs and to make changes, but those who do not feel confident about making major changes will be able to do all the main computations by changing only one function definition.

Because of the emphasis on qualitative aspects and the flexibility with regard to program modification, it is hoped that the book will prove useful

1

to a wide variety of students. A certain basis of mathematical knowledge must be assumed; the text will not prove easily understandable to the general reader. But science or mathematics students preparing for university entrance or in the first year at university (or those with equivalent knowledge) ought to be able to follow the earlier sections and derive some benefit from the more advanced sections that follow. The text has been written with college students in mind, expecting only a knowledge of elementary calculus, trigonometric functions, and complex numbers.

The prerequisites for each part are:

Part 1: Idea of functions
          Cartesian coordinates and axes
          Graphs of functions

Part 2: Trigonometric functions cos, sin

Part 3: Complex numbers
          Calculus: elementary integrals.

$\rangle$ Part 1

$\rangle$ Sampling and Resolution

Readers who are familiar with the concepts of sampling and resolution should skip to Part 2.

A request like 'draw a graph of $y = x^2$' sounds simple enough but turns out to be a little vague on further reflection. What values of x should be used? What scales? How accurately should the job be done? The human mind is very good at coping with this kind of vagueness; if no further information is supplied, most people familiar with this kind of mathematical task would draw a rough pair of Cartesian axes on a plain piece of paper and then a free-hand parabola; probably they would ensure the parabola went through the origin, touching the x axis there. They can do this because at some time they have plotted the function properly, or seen someone else do it, and have remembered the essential features.

To do the task properly, we must first agree on a range of values of x, for example -10 to $+10$. In principle we must next calculate $y = x^2$ for every value of x in this range; as there are an infinite number of values this is impossible, so instead a finite number of in-between values are chosen, for example the 21 values $x = -10, -9, -8,...,0, 1,...,10$. Next we work out $y = x^2$ at each of these values and assume that these y values will be typical of any further values we may eventually obtain. On this assumption we can then take a piece of graph paper, decide where to rule the axes, and what scales to use. We can now mark a dot for each pair of values (x, y).

The values of y obtained in this way are called sample values of $y = x^2$ obtained at the sample points x. There is no need for the sample points to be regularly spaced; we might look at our graph paper and decide that some points are rather far apart, and we would like some values in-between. The usual procedure for completing our graph would be to add more sample points until they looked close enough together to join each neighbouring dot with a straight pencil line. This is another vague

3

requirement which could be resolved by asking, for example, that 'no point on the line should be vertically displaced by more than 0.1 units of y from the true value for that x'. Even now there is still some vagueness because every pencil line has finite thickness.

For the purpose of plotting our function, it is becoming clear that whatever we do, we can never plot it exactly. We had better give up the idea and accept that there will always be some fuzziness in our efforts.

'Fuzziness' in plotting can be illustrated by a simple computer graphics demonstration. All common computer screens work by illuminating individual blobs on the screen: these blobs cannot be subdivided. Our rule for plotting any function will be: sample the function at suitably spaced values of x, and illuminate the blob which covers the resulting point (x, y). If the blobs are fine enough and the sample points are sufficiently closely spaced, our eyes and brain will see them as a continuous line, if we don't look too closely. The finer the blobs, the better we will say is the *screen resolution*; the closer the sample points, the better we will say is the *sample resolution*.

These effects are illustrated by the program SAMPLE1 (see separate description). As supplied it will plot $y = x^2$ in the range $-10 \leqslant x \leqslant 10$ using a screen resolution which you input when the program asks. The figure required is the number of blobs per full screen width: you may not ask for fewer than 20, but there is no upper limit; if you ask for greater resolution than actually possessed by the screen, the program will take longer to run but you will not end up with a better figure. Notice the order in which the blobs are illuminated: at first the function is sampled at widely spaced

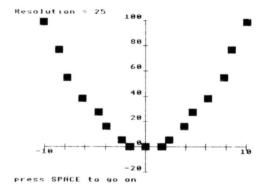

Figure 1.1   Typical display from SAMPLE1, showing a partially sampled function at low resolution.

points, and then the values in between are filled in. If you ask for lower resolution than actually possessed by the screen, a square region of blobs will be illuminated to simulate that lower resolution.

Run SAMPLE1 a few times with various resolutions. Convince yourself that it is pointless to go beyond maximum screen resolution, and note the number of function evaluations required at maximum resolution.

Convince yourself also that you do not need to go to full resolution to get a very good idea of how the graph looks. If you are confident with the computer, try changing the program to graph some other simple functions; for example, $y = e^x$, $y = \sqrt{x}$, $y = \sin x$, etc. Remember to choose suitable ranges of x and y.

You will have noted the time needed to plot at full resolution, and perhaps you are wondering if we can make an acceptable graph out of fewer points by joining them up with lines, just as one would do with pencil and ruler on graph paper? The computer can do this too, as demonstrated in program SAMPLE2. This time when you run the program you must input a number which will be the number of sample points at which the computer will evaluate the function. The resulting points will be joined with lines. The computer will then offer you the options of repeating this display with a new value for the number of sample points, or continuing, in which case the function is plotted at the finest possible screen resolution (as for SAMPLE1), and then the graph resulting from the line plotting method is superimposed (in a different colour if you are using colour graphics). From this display you can see

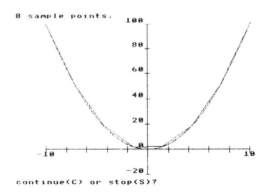

Figure 1.2 Typical display from SAMPLE2, showing an approximation made of straight line segments superimposed on the actual function plotted at the highest available resolution.

where the line plotting method goes wrong, because if the line plotting was as accurate as the screen resolution allows, all of the point plot would be covered over by the line plot. Experiment with the program, starting with two sample points and increasing the number up to the limit of screen resolution. Observe how a quite acceptable graph can be obtained with as few as 20 sample points, and that eventually the line plot and the point plot become identical; but the most important point to note is that the line plot becomes very close indeed to the point plot when the sample resolution is considerably lower than the screen resolution. Later you will see that this is not true under all conditions; the nature of the function and the range over which it is to be plotted affect matters a great deal. If you are using a colour display, you will notice that even when the two plots are getting very close (say beyond 40 sample points) a few isolated spots show up as different. This is due to computer rounding errors.

When we, or the computer, join up neighbouring points with lines, we are in effect guessing the y values at all the x values between points instead of working them out directly. We are replacing the true graph with the graph of a function which is linear, meaning simply that it has a straight line graph. A different linear function is used between each pair of points. There is a mathematical phrase for this: *linear interpolation*. Interpolation means guessing the in-between values; linear means use straight lines to do the guessing. There are more complicated ways of doing interpolation, but we will not discuss them here.

What should be concluded from these demonstrations? In every practical application of mathematics, there will be some limit of resolution imposed by the real world; for example, the graininess of a screen, the tolerance with which a metal component can be machined, the limit of accuracy of some measuring device like a stop-watch or weighing scales, etc. So a totally accurate representation of a mathematical function will never be needed; there is just no purpose in attempting to obtain accuracy beyond the limiting resolution. Usually a lower resolution will be adequate for the purpose, which can be chosen according to circumstances, but whatever the resolution that is set, there will be a sampling resolution that combined with interpolation will enable the values in question to be represented with acceptable accuracy.

Mathematical notation:

Sample points will be denoted by $x_i : i = 1, 2, ..., N$.
Sample values will be denoted by $y_i : i = 1, 2, ..., N$.

Each sample point would be plotted at coordinates (x_i, y_i).

〉 Part 2

〉 Periodic Functions, Harmonics and Fourier Series

〉2.1 Periodicity

Periodic functions are functions which repeat themselves over and over again. They are very important in many applications of mathematics, for example in biological modelling to represent the daily cycle of light and dark, in mechanical engineering to represent the repeated stresses and strains of a rotating component, in the theory of radio waves and circuits, and in acoustics, where a sound wave is due to constantly repeated motions of the molecules of air. Because all these examples involve the idea of time, it is natural to use the symbol t, rather than x, to stand for the independent variable; that is, we will talk of functions $y = t^2$ (for example) rather than $y = x^2$. When plotting graphs, t will be measured along the horizontal axis.

As a simple example of a periodic function, take

$$y = \cos(2\pi t).$$

You can obtain a graph of this function by using the program FSCOS. Refer to the operating guide for more details of how to run this program, but note that for this example you need only reply to each question by pressing the RETURN key, which will select the values that appear in brackets. The prompts that you should see are as follows:

Const $(=0):?$

Up to what frequency multiple of 2.PI $(=1)?$

1: Coeff. of cos (2.PI.t) $(=1)?$

7

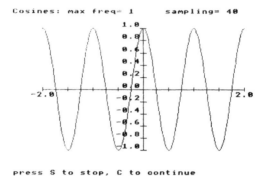

press S to stop, C to continue

Figure 2.1 Display obtained from FSCOS using default data values, as suggested in §2.1.

How many periods to plot ($=4$)?

Sampling rate ($=40$)?

The meaning of these values will become clear later. For the moment just press RETURN in reply to each prompt.

You should then get a graph with portions plotted in different styles (solid, dotted or different colours). Each portion represents one period; if you took any portion and shifted it along the t axis a whole number of units, it would coincide exactly with the values already plotted there. This is what is meant by saying that the function is periodic. What is more, the portions do not need to have any particular starting point. You could take any portion of any length, and if you shifted it a whole number of units it would coincide with the values already there. The minimum shift with this property (apart from the obviously trivial case of zero shift) is one unit in this example, and we say that $y=\cos(2\pi t)$ is periodic with period $=1$. The period is defined to be the minimum shift because there are other, larger shifts with the same property; in fact any whole number multiple of the period will do.

$\rangle$2.2 Harmonics

Consider now the function $y=\cos(4\pi t)$. If you are familiar with trig functions you may not need to use the computer to see that this function is also periodic, this time with a period of $\frac{1}{2}$, but if you like you can use

FSCOS to get a graph. The dialogue needed is indicated below. You can reply with a number, ending with RETURN, or just press RETURN, which is equivalent to typing the number in brackets.

Const (=0)?

Up to what frequency multiple of 2.PI (=1)? 2

1: Coeff. of cos (2.PI.t) (=1)? 0

2: Coeff. of cos (4.PI.t) (=0)? 1

How many periods to plot (=4)?

Sampling rate (=40)?

You will find that the function now goes through eight cycles, as compared with four on the previous demonstration. If you now understand the meaning of period, you will be thinking that this function has gone through eight periods, not four as you specified in the data. This is because the period is being measured in terms of the period of $y = \cos(2\pi t)$; in other words, in the time it takes $\cos(2\pi t)$ to go through four cycles, $\cos(4\pi t)$ goes through eight. The period of $\cos(4\pi t)$ is $\frac{1}{2}$.

It is time to name some names: the function $\cos(2\pi t)$ will be called the fundamental, $\cos(4\pi t)$ the first harmonic, $\cos(6\pi t)$ the second harmonic, etc. If n is a whole number, as printed in front of the coefficient prompt, then $\cos(2n\pi t)$ is the $(n-1)$th harmonic. So the fundamental is the same as the zeroth harmonic. Our unit of time is the fundamental period, and the period of the $(n-1)$th harmonic is $1/n$. It is now a good moment also to

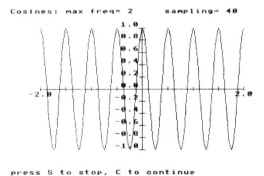

Cosines: max freq= 2   sampling= 40

press S to stop, C to continue

Figure 2.2 Display obtained from FSCOS using modified data values as suggested in §2.2.

define frequency; using cycles per unit time as our measure, the frequency of the $(n-1)$th harmonic is n.

The computer illustrations suggested so far have been for the fundamental, with period 1 and frequency 1, and the first harmonic, with frequency 2 and period $\frac{1}{2}$. You may like to try out program FSCOS for higher harmonics; make all the coefficients zero except for the harmonic you want to look at. The higher the frequency, the more rapidly the function oscillates, and the shorter is its period. Use a sampling rate of $10n$.

It was remarked a short while back that if any section of a periodic function is shifted sideways along the t axis by a whole number of periods then that section will be coincident with the curve. Now for the first and higher harmonics, it will always be true that a shift equal to one fundamental period will satisfy this condition because, as stated, the period of the $(n-1)$th harmonic is $1/n$; therefore multiplication by the whole number n will yield the value 1, equal to the fundamental period. It should now be seen that a composite fundamental and at least one other harmonic will itself be periodic at the fundamental period. For example, consider

$$y = \cos(2\pi t) + \cos(4\pi t).$$

It could be said that this function consists of *equal amounts of the fundamental and the first harmonic*. This is obviously periodic; but why is the period equal to 1 and not $\frac{1}{2}$? The answer clearly is that when the first harmonic returns to its starting value at $t = \frac{1}{2}$ the fundamental has only got half way through its cycle, so in this case the slowest oscillation determines the period. This is why program FSCOS sticks to using the fundamental period as its unit of time.

Program FSCOS allows you to visualise the effect of combining up to 32 harmonics in this way. The *coefficients* required by the program determine the relative amounts of each harmonic to be added; in the above example, these coefficients are both unity. Experiment with other combinations, and watch the following points.

(a) The *sampling rate* determines the number of sample points per period. The interval between sample points is 1 divided by the sampling rate, so that if the endpoints of each period are counted there are actually $(1 + \text{sampling rate})$ points per period.

(b) There is no need for the fundamental to be present; in this case the frequency of the resulting function is the highest common factor of the frequencies present.

(c) If you use the higher frequencies, you may have to be content with plotting just one or two periods on the screen to see all the detail clearly.

(d) The effect of the constant (the first data item) is merely to add a constant value to the function. Note that it does not affect the period, and that it can be thought of as a harmonic with zero frequency. Try changing the constants given in the examples.

Here is a summary of values suggested so far for FSCOS, plus some other values to try, in the order:

constant, maximum frequency, coefficients in order.

It is left to the reader to choose values for number of periods and sampling rate. Note that the program allows input in the form of rational fractions.

Example No.	Constant	Maximum frequency	Coefficients
1	0	1	1
2	0	2	0,1
3	0	2	1,1
4	1.2	2	1,1
5	0	2	1,2
6	0	5	1,0,0,0,1
7	0	10	0,0,0,0,0,0,0,0,1,1
8	0	7	1,1/2,1/4,1/8,1/16,1/32,1/64
9	0	7	$1,-1/2,1/4,-1/8,1/16,-1/32,1/64$
10	0	5	$1,-1/2,1/6,-1/24,1/120$

⟩2.3 Symmetry and parity

Quite a wide variety of periodic functions can evidently be generated in this way. This suggests a question of central importance to the theory known as *Fourier analysis*: could any periodic function be generated in this way? As things stand, the answer must be no, because — you may have noticed — all the functions generated by FSCOS have a property that a general periodic function does not have, namely symmetry. If you didn't spot it, look again at some displays and notice that the function value at any value of t is exactly the same as the function value at $-t$; if you think

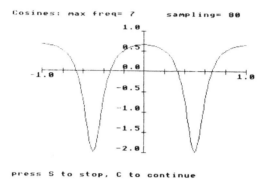

Figure 2.3   Display obtained from FSCOS using data from example 9, §2.2.

of the screen as a piece of paper, folding it along the y axis would bring the two portions of the graph on opposite sides of the axis together. Functions with this kind of symmetry are said to be even, or to have even parity. You can also see that folding along a vertical line through *any* integer value of t produces similar results, because the function is even *and* periodic.

There is another kind of symmetry called odd parity; for example, the function

$$y = \sin(2\pi t)$$

is an odd function. You can run the program FSSIN to see it graphed; the program omits the constant but otherwise runs exactly like FSCOS, using sine functions everywhere that cosines were used before. It is interesting to repeat all the examples that were suggested for FSCOS. Observe that for odd parity functions, the value corresponding to any value of t is exactly minus one times the value at $-t$; functions with this property must therefore take the value zero at $t = 0$ (why?). From now on, the cosine functions that we used may also be called *even harmonics*, and the corresponding sine functions *odd harmonics*. Can you also see why the constant was included in FSCOS but not in FSSIN?

Returning to our central question, we see that this must now be rephrased to: can any even function be represented by a combination of even harmonics, and any odd function by a similar combination of odd harmonics? And we must add a supplementary question, what about more general functions?

Answering the last question first, it is not too difficult to see that a general function (not necessarily harmonic, either) may be represented as the combination of an odd function and an even function. If $F(t)$ is a general function, let $G(t)$, $H(t)$ be two other functions defined by

$$G(t) = \tfrac{1}{2}(F(t) + F(-t))$$

$$H(t) = \tfrac{1}{2}(F(t) - F(-t)).$$

It is immediate that $F(t) = G(t) + H(t)$. Prove for yourself that $G(t)$ is even, and $H(t)$ is odd.

$\rangle$2.4 Lack of smoothness

With this result, the conjecture that any periodic function can be represented by a combination of harmonics becomes more plausible. The first step in finding such a combination is to find the even and odd functions that make it up. As this is a simple step, the next few examples will be purely even or purely odd functions. But there is obviously another property that any combination of harmonics must have. It will have to be smooth, because each harmonic is smooth — or will it? Try out the following examples, repeating each set of data for each sampling rate shown.

Example No.	Constant	Maximum frequency	Coefficients	No. of periods	Sampling rate
11	0	3	1,0,1/9	2	5,10,20
12	0	5	1,0,1/9,0,1/25	2	10,20,40
13	0	7	1,0,1/9,0,1/25,0,1/49	2	20,40,80
14	0	9	1,0,1/9,0,1/25,0,1/49 0,1/81	1	20,40,80,100

What you should observe is that the apparent smoothness of the graphs depends very much on the number of sample points. For any given combination of harmonics, the graphs get smoother with more sample points, as only to be expected. But the more harmonics included, the more sharply the function seems to vary, or change course, at a few places, whilst remaining smooth everywhere else. Places like this where rapid changes occur represent *discontinuities* in the function; or rather we should say *appear* to represent discontinuities because we know that if we use a finite

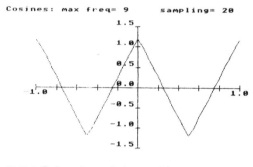

Figure 2.4 Display obtained from FSCOS using data from example 14, §2.4.

number of harmonics and increase the sampling rate enough, we will see a smooth curve. However in any actual application as we have noted before there is a natural limit to resolution and therefore to sampling rate, so a plausible conjecture suggests itself: we can represent, at any required resolution, a periodic function with a finite number of actual discontinuities by using an appropriate series and going up to high enough frequencies.

$\rangle$2.5 Sampling rate

There is a critical value for the sampling rate (by which is meant the number of sample points per fundamental period). This critical value is related to the highest harmonic used, and it is very easy to see how it arises. Use FSCOS to produce a graph of the first even harmonic (frequency 2), plotting over just one fundamental period. Start plotting with 32 sample points per period, and then repeat with 16, 8, 4, 2 and 1. You should have noticed a drastic change comparing the results at sampling rates 16, 8, 4 or 2: 16 should produce a recognisable harmonic curve, even though it is visibly made up of lines; 8 and 4 should produce a *sawtooth* shape, and 2 a flat line. If you now go back to the curve drawn with 16 sample points per period, you can see what is happening. At a sampling rate of 4, the sample points coincide exactly with successive peaks and troughs, hence the sawtooth. Halving the sampling rate means the troughs are skipped, giving a totally false picture of the harmonic. The critical value of sampling rate is 4, exactly twice the frequency of the harmonic itself; anything less will miss out the important landmarks of the function. Work out for yourself, with a freehand sketch, where the sample points occur at sampling rates 8 and 16.

Two points need to be settled about critical sampling rate. First, if you repeat the above work using FSSIN to plot the third odd harmonic, you will observe the sawtooth waveform occurring at twice the critical sampling rate. Inspection of the function at a higher sampling rate should show you the reason for this. The second point is, why haven't any sampling rates been suggested at other intermediate values? Try some out on your computer (for example, 3, 5, 6) and you will see that even more misleading graphs result; these effects, particularly when associated with sample rates near or below the critical sampling rate, are known as *aliasing*. We obviously want the sampling period to be a whole number subdivision of the period of the harmonic, which means the sampling rate should be a multiple of the harmonic frequency. Given this, the critical sampling rate is the lowest frequency that can detect the presence of a harmonic at the frequency in question. (But note, when several harmonics are combined, the critical frequency of the highest harmonic may not be a multiple of all of the other, lower harmonic frequencies.)

Returning to the conjecture at the end of §2.4 about representing discontinuities, or jerks, it is obvious that the sampling rate must be related to the size of the t intervals to which the jerks are to be confined. The smaller these intervals, the higher the sampling rate will have to be. In Part 1, the proposition was put forward that there is always some limit of resolution beyond which it is either impossible to go, or one does not wish to go. It looks as if besides smooth functions, it will be possible to reproduce periodic functions with jerks, provided we do not look closely at what happens within the jerk region. That means that the sampling interval must be set equal to this limit of resolution. Having settled the sampling rate, it is obviously sensible to limit the harmonics to be used so that the highest frequency is half the sampling rate. Any higher frequencies will either 'slip through' the sample points or produce misleading values, as the last computer run showed.

⟩2.6 Fourier series

Now let us look back over our attempts to represent any given periodic function by a combination of harmonics.

1.   The period of the fundamental must be taken as the same as the period

of the given function, otherwise no combination of harmonics can ever reproduce the correct period.

2. The given function may be split into even and odd components.

3. A sampling interval must be set to determine within what resolution the given function is to be reproduced, or equivalently a sampling rate chosen.

   Note: sampling interval = period divided by sampling rate per period.

4. Half the sampling rate is the upper limit of harmonic frequencies to be used.

This is as far as we have got. The next step, step 5, is the hardest: determine the coefficients so that the values of the given function are correctly reproduced at the sample points. Indeed, the question 'can it be done?' has not yet been answered.

Computer demonstrations can never *prove* an assertion of this nature, only help to make it plausible and provide some insight into *how* and *why*. It has been demonstrated that functions with some kinds of sharp changes in form can be reproduced up to the resolution limit, making it all the more plausible that smooth functions can be represented in this way. In fact the proof of this assertion, and the determination of the precise conditions under which it holds, use some very advanced mathematics, and here we must be content with knowing that rigorous mathematical proof does exist. For our purposes it can be stated that provided a given periodic function has finite values, and at most a finite number of jerks of the sort demonstrated, then a Fourier series can be found which gives the coefficients of the harmonics required. The series is infinite; that is, we can obtain values from it for as many harmonics as we like; the more that are used, the finer the resolution with which the function will be reproduced, or synthesised. The reverse process, that is starting with a given function and obtaining the coefficients, is called Fourier analysis: more about how this may be done will be found in the next section.

⟩2.7 An application

What use, you may wonder, is this knowledge? It turns out to be extremely useful in many different areas, because it is often easier to understand how a complicated system, for example a radio circuit, responds to a pure harmonic input rather than to a more realistic input. Fourier analysis enables a realistic input to be Fourier analysed into harmonics; if it is known what the circuit does to each harmonic separately then the output will be the synthesis of each processed harmonic. The musical instrument known as a synthesiser uses this principle by allowing the musician to combine harmonics at will. Many of the computer runs that you have done have been a visual representation of this.

As an example, consider the repeated on–off function that could be obtained by operating a switch at regular intervals of time. There is a Fourier series for this, but it requires many more harmonics than any of the examples so far used. Run FSCOS using the following values to make a plausible case that a Fourier series can achieve this.

Example No.	Constant	Maximum frequency	Coefficients	No. of periods	Sampling rate
15	0	8	$1,0,-1/3,0,1/5,0,-1/7,0$	2	16
16	0	16	continue up to $-1/15,0$	2	32
17	0	32	continue up to $-1/31,0$	2	64

To achieve a really convincing result, very much higher harmonics must be used. But notice with these results how a relatively smooth area appears to be forming mid-way between the jump regions, and the jumps themselves appear to get steeper as more harmonics and sample points are used. Repeat the examples, displaying just one period to see this better.

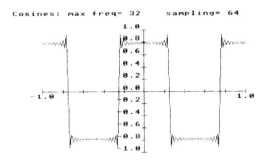

Figure 2.5 Display obtained from FSCOS using data from example 17, §2.7; on–off function truncated at frequency 32 times the fundamental.

Now suppose that this *on–off* or *square wave* function represents current flowing in a light circuit, the y axis measuring current. The exact Fourier series is the infinite series

$$1,0,-1/3,0,1/5,0,-1/7,0,1/9,0,-1/11,0,1/13,\ldots.$$

The decision where to cut it depends on the resolution to which the function is to be measured. Suppose we are interested in why the operation of light switches causes interference with radios or TVs: as these can detect frequencies in the 500 kHz to 1000 MHz range (half a million to one billion cycles per second), and assuming the switch is operated about once a second, the series will have to be taken to between the 500 000th to billionth term, because the radio set can sample electrical fields at this sort of frequency. Some of the harmonics will be close to the frequency at which the set is tuned, and hence interference will result. Of course the strength of the harmonic must be enough to be detectable, and the series shows that the strength should decrease with increasing frequency, but it decreases only slowly. This also explains why switches affect radio reception over a very wide frequency range.

Other shapes of current flow, for instance examples 11–14, have Fourier coefficients which decay much more rapidly:

$$1,0,1/9,0,1/25,0,\ldots,1/n^2 \quad (n \text{ odd}),\ldots.$$

The strength of the 1 MHz harmonic would now be 1 divided by $(1000\,000)^2$, one millionth of the strength of the harmonic produced by the switch. In fact there is a generalisation of this result: the more sudden the type of change in function values, the more slowly the harmonic strengths decay.

You can also see the effect of an electronic filter on current surges of this sort. These are devices which will not allow frequencies above a certain limit to flow through them. To simplify matters, it will be supposed that a filter is available which allows all frequencies below a certain value to flow without modification, but cuts off all above this value. If, for example, the cut-off frequency was 8, the effect of the filter on the periodic on–off switch would be to produce a current flow proportional to

$$\cos(2\pi t) - \frac{1}{3}\cos(6\pi t) + \frac{1}{5}\cos(10\pi t) - \frac{1}{7}\cos(14\pi t).$$

From earlier remarks, you might be tempted to think that the appropriate sampling frequency would be 14, but this would be incorrect because this time we are saying that this function is the *actual* output from the filter, not

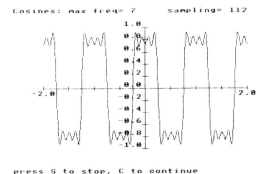

Figure 2.6   The effect of a filter on an alternating on–off signal, discussed in §2.7.

an approximation to it. The sampling frequency needs to be high enough to resolve the highest remaining harmonic as a smooth curve.

It might be objected that in practice people do not turn switches on and off at precise one second intervals, so how can this analysis be meaningful? The answer is that we have oversimplified, but it remains true that rapid changes, even if not periodic, give rise to high harmonics. There will be some further explanation in Part 3.

⟩2.8   Mathematical notation and the general Fourier series

Let $F(t)$ be a periodic function, smooth except for a finite number of jerks or jumps of the sort described earlier, and having period P. No kind of symmetry need be assumed. The Fourier series for this function, up to the Nth frequency, will be

$$F_N(t) = a_o + \sum_{n=1}^{N} a_n \cos\left(\frac{2\pi nt}{P}\right) + \sum_{n=1}^{N} b_n \sin\left(\frac{2\pi nt}{P}\right)$$

where $a_o, a_1, ..., b_1, b_2, ...$ are constant coefficients. Some series of interest (using $P = 1$ throughout) are given below.

(i)   The on–off function, as in example 15.

$$|t| < \tfrac{1}{4}: F(t) = \pi/2$$

$$\tfrac{1}{4} < |t| < \tfrac{1}{2}: F(t) = 0$$

series:

$$a_o = \frac{\pi}{4}, \quad a_{2n-1} = \frac{(-1)^n}{2n-1}, \quad a_{2n} = 0 \qquad n = 0,1,2,...$$

$$b_n = 0 \qquad n = 1,2,....$$

(ii)   The triangle function, as in example 11.

$$-\tfrac{1}{2} < t < 0 : F(t) = \frac{\pi^2}{4}(2t+1)$$

$$0 < t < \tfrac{1}{2} : F(t) = \frac{\pi^2}{4}(1-2t)$$

series:

$$a_0 = \frac{\pi^2}{8}, \quad a_{2n-1} = \frac{1}{(2n-1)^2}, \quad a_{2n} = 0 \qquad n = 0,1,2,....$$

$$b_n = 0 \qquad n = 1,2,3,....$$

(iii)   The sawtooth function.

$$|t| < \tfrac{1}{2} : F(t) = \pi t$$

series:

$$a_n = 0 \qquad n = 0,1,2,...$$

$$b_n = \frac{(-1)^{n+1}}{n} \qquad n = 1,2,3,....$$

(iv)   The on–off function with odd parity.

$$0 < t < \tfrac{1}{2} : F(t) = \pi/4$$

$$-\tfrac{1}{2} < t < 0 : F(t) = -\pi/4$$

series:

$$a_n = 0 \qquad n = 0,1,2,...$$

$$b_{2n+1} = \frac{1}{2n+1}, \quad b_{2n} = 0 \qquad n = 0,1,2,....$$

$\rangle$2.9   Combined series and further examples

With the aid of the formulae in the previous section, it is possible to obtain series for similar shapes at different frequencies. For example, the odd

parity on–off function at three times the fundamental frequency can be
obtained by choosing the period $P = \frac{1}{3}$ and using the previous coefficients
to give

$$F_N(t) = \sum_{n=1}^{N} b_n \sin(6\pi n t)$$

where even numbered b_n are zero, odd numbered b_n are $1/n$, i.e.

$$\sin(6\pi t) + \frac{1}{3}\sin(18\pi t) + \frac{1}{5}\sin(30\pi t) + \ldots$$

The whole series can be multiplied by an overall constant, of course, to
change its amplitude without changing its shape. Notice that only those
frequencies are used which are multiples of 3, because a new fundamental
period is being used.

A combination of such a function and one at the original frequency will
still be periodic at the original frequency, provided a whole number of new
fundamental periods fit into the old. For example, the Fourier sine series
for the combination of this new fundamental (at half the original
amplitude) and the sawtooth would be:

$$1, \ -\tfrac{1}{2}, \ \tfrac{1}{3} + \tfrac{1}{2}(1), \ -\tfrac{1}{4}, \ \tfrac{1}{5}, \ -\tfrac{1}{6}, \ \tfrac{1}{7}, \ -\tfrac{1}{8}, \ \tfrac{1}{9} + \tfrac{1}{2}(\tfrac{1}{3}), \ \tfrac{1}{10}, \ \tfrac{1}{11}, \ -\tfrac{1}{12}, \ \tfrac{1}{13}, \ -\tfrac{1}{14}, \ \tfrac{1}{15} + \tfrac{1}{2}(\tfrac{1}{5}), \ldots$$

Combined sine and cosine series can be generated in a similar way. The
program FSBOTH will do this.

Experiment with FSBOTH, combining some of the series already
introduced, and varying their amplitude.

〉 Part 3

〉 Fourier Transforms

〉**3.1   Introduction**

In part 2, the idea was developed that a periodic function whose values were measured at a number of sample points could be represented by a limited number of Fourier harmonics (with carefully chosen coefficients) to whatever accuracy was required. The essential point was that the sampling imposed a limit of resolution, beyond which it was pointless to go, and that this in turn imposed a frequency limit which it was pointless to exceed. However, any claim that the sample values could be matched exactly by this limited number of harmonics was carefully avoided.

But if we look at the problem from an information theory point of view, we might expect that the data content of knowing the strengths of N harmonics should be the same as knowing N sample values. After all, we are free to choose the coefficients (amplitudes) of N harmonics, so couldn't we choose them in such a way that the sample values are reached exactly? This expectation can indeed be realised, and the resulting set of values are known as the *discrete Fourier transform* (DFT) of the original sample values.

〉**3.2   An example**

In order to make more progress with this idea, it will be essential to use more mathematical notation than we have done so far. But before coming to this, program FOURIER can be used as an illustration. FOURIER as supplied will use the function

22

$$f(t) = \begin{cases} 1 & -1 < t < 1 \\ \frac{1}{2} & t = +1 \text{ or } t = -1 \\ 0 & \text{otherwise.} \end{cases} \qquad (3.1)$$

The reason for defining $f(t)$ to have the value $\frac{1}{2}$ at $+1$ or -1 will be given later. Although as defined here $f(t)$ is not periodic, the program actually works with a periodic function derived from $f(t)$ in the following simple way. As part of the data input to the program you will be expected to provide a period, T, which will be called the *imposed period*. The program will then calculate $f(t)$ at sample points only within the range

$$-\tfrac{1}{2}T \leqslant t < \tfrac{1}{2}T.$$

Outside this range, the function values are simply repeated periodically. Let us call this function $F(t)$; it is by definition periodic, and within the range it is exactly equal to $f(t)$. Notice that $F(t)$ is an even periodic function, and that it consists of jumps, provided $T > 2$ (if $T < 2$ then $f(t)$ will only ever be calculated in the region where $f(t) = 1$). Notice also that $F(t)$ is very like the function towards which we were moving in examples 15 to 17 in Part 2. Taking $T = 4$ will generate an $F(t)$ which has period 4, and spends equal amounts of time at values 0 and 1, apart from the spot value of $\frac{1}{2}$ at $t = +1$. The Part 2 example never got to be quite flat, of course, but it did alternate, between 'up' and 'down' in equal amounts. The 'down' value could obviously be raised to nearly zero by adding on a constant, which as we saw could be regarded as a trivial first term of a Fourier series. The length of the period is obviously just a matter of units and would not be expected to change the nature of the results.

Now run FOURIER. The program will prompt you for two data items: power of two and range of t. The second number is just the value of T that we have been talking about. The first determines the number of sample points that will be used; if M is the power of 2, then 2^M sample points will be used. Start with $M = 3$, $T = 4$. You should obtain two graphs. If you have a colour monitor you will see that each consists of two separate plots, but as one lies along the horizontal axis it will not show up on a monochrome monitor. In either case ignore it for now. The top graph shows the function $F(t)$, but it will not look exactly as expected as there will be a sloping line in the region $\frac{1}{2} < t < \frac{3}{2}$, whereas $F(t)$ was expected to jump from 1 to 0 suddenly across the value $t = 1$. This ramp is of course due to the spacing out of the sample points; for $M = 3$ there are eight sample points per period which makes the spacing equal to $\frac{1}{2}$. The ramp runs across two sample intervals, and will get steeper as the sampling rate increases. Look now at the lower graph (the DFT). It too is made up of eight sample intervals (nine points including the ends); its horizontal axis measures frequency, and its vertical axis measures the value of the Fourier coefficients. For the moment ignore the negative frequency part of the graph, and notice that the positive part is made up of four straight lines. Only the ends of these lines have any significance for Fourier series; the frequency values are 0, 0.25, 0.5, 0.75, and 1. In the previous section, units of time were chosen so that the period of the periodic function was 1; in this section, that is no longer necessarily the case. The period of $F(t)$ was chosen to be 4; its frequency is therefore $\frac{1}{4}$, the number of complete cycles per unit time; note that in these units frequency times period equals 1. The fundamental of $F(t)$ will therefore also have frquency $\frac{1}{4}$, and it should now be clear that the values 0, 0.25, 0.5, 0.75, and 1 correspond to the first four harmonics of the Fourier series for $F(t)$, together with the zero frequency constant. The sampling rate for $F(t)$ was 2 (period $= 4$, therefore eight points per period); from the discussion in the previous section, you should expect the maximum frequency of harmonics to be half this, namely 1. This is consistent with using only the first four harmonics (plus the constant). Look now at the heights corresponding to these frequencies. The coefficients that were used in example 15 in Part 2 were:

$$\text{const} = 0, \quad \text{harmonic coeffs} = 1, 0, -\frac{1}{3}, 0, \frac{1}{5}, \dots \qquad (3.2a)$$

The function produced by this series, theoretically, alternates between values $\pi/4$ and $-\pi/4$, and program FSCOS verified this approximately. To produce $F(t)$, the series will have to be scaled up by a factor $2/\pi$ and a constant of $\frac{1}{2}$ added on, giving the modified Fourier series

$$\text{const} = \tfrac{1}{2}, \quad \text{harmonic coeffs} = \frac{2}{\pi}\left(1, 0, -\frac{1}{3}, 0, \frac{1}{5}, \dots\right). \qquad (3.2b)$$

For reasons that will emerge later, we want now to compare this series with the DFT divided by T ($= 4$ in this example). Looking at the graph, the value at zero frequency is 2, and $2/T = \frac{1}{2}$, which agrees with the constant in the modified Fourier series. It is also clear that the zero values at frequencies 0.5 and 1 match the zeros in the Fourier series. But the values

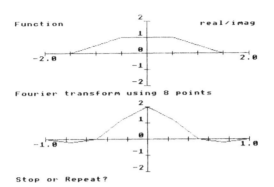

Figure 3.1   First suggested run of FOURIER using $M=3$, $T=4$.

at the odd harmonic frequencies 0.25 and 0.75 look to be about 1.2 and -0.2; the change of sign agrees with the Fourier series, but after dividing by T we get values of about 0.3 and -0.05 whereas $2/\pi = -0.2$ (approximately). It looks as if there is a factor of 2 missing for the first value, and a factor of 4 for the second. To account for the factor of two, the values at negative frequency must be included (a justification will be given later); this is done simply by regarding the values at the corresponding negative frequency as an additional contribution to the Fourier series, ignoring the sign of the frequency. This will not affect the central, zero frequency value, because it has no separate negative frequency counterpart. To get better agreement for the next non-zero value, it will be necessary to use more sample points.

Run FOURIER again, using $M=4$, $T=4$. This time there will be 16 sample points per period, so the sampling rate has been doubled. Notice that the frequency range of the lower graph is doubled, and the heights (divided by T) continue in approximate agreement with the Fourier series. Continue to run FOURIER, keeping $T=4$ but increasing M by 1 each time (so doubling the sampling rate) until the size is too big for your computer to handle (you will get some kind of error message). Notice how in the top graph the ramps get steeper, and how the alternating nature of the discrete Fourier transform continues to agree (very approximately) with the Fourier series for $F(t)$.

We must now examine how closely the discrete Fourier transform (DFT) agrees with the Fourier series. To do this, it is best not to use graphics but to print out actual values. Before running the program that

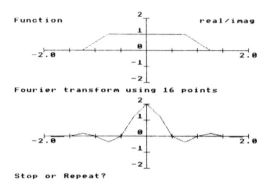

Figure 3.2   Second suggested run of FOURIER using $M=4$, $T=4$.

does this, the mathematical notation for what we are doing must be developed.

⟩3.3   Using complex numbers

To develop the theory of discrete Fourier transforms (DFT) we will need to use some notation from the theory of complex numbers. The central relation that we will need is

$$\exp(iA) = \cos(A) + i\,\sin(A) \qquad (3.3)$$

where i is the square root of -1, A is any real number, thought of as an angle and often called the argument of the complex number represented by either side of (3.3). By inspection, taking $A=0$ gives $\exp(iA)=1$; $A=\pi/2$ (radians) gives $\exp(i\pi/2)=i$; and $A=\pi$ gives $\exp(i\pi)=-1$.

The general cosine Fourier series with fundamental frequency 1 may be written

$$\mathrm{Re}\left(\sum_{r=0}^{\infty} a_r \exp(2\pi i r)\right) \qquad (3.4)$$

where the coefficients a_r are real numbers, and the general sine Fourier series may be written

$$\mathrm{Im}\left(\sum_{r=1}^{\infty} b_r \exp(2\pi i r)\right) \qquad (3.5)$$

where the coefficients b_r are real numbers.

Complex number notation makes it just as easy to deal with a general sine and cosine Fourier series combined by allowing the coefficients to be complex. Take

$$\frac{1}{T} C_r = a_r - i\, b_r, \quad r = 0,1,2,\ldots$$

and consider

$$\frac{1}{T} \sum_{r=0}^{\infty} C_r \exp(2\pi i\, rt) \tag{3.6}$$

(the reason for inserting the factor $1/T$ will become clear later).

Each term is in the form

$$(a_r - ib_r)[\cos(2\pi rt) + i\, \sin(2\pi rt)]$$
$$= [a_r \cos(2\pi rt) + b_r \sin(2\pi rt)] + i[a_r \sin(2\pi rt) - b_r \cos(2\pi rt)].$$

So the real part of (3.6) is in the form of a general Fourier series, and notice that b_0 does not have to be defined because $\sin 2\pi.0 = 0$. The imaginary part of (3.6) is also a Fourier series, namely, the series obtained from the previous one by switching the coefficients in the sine and cosine parts. The relationship between the two series is of interest but for the moment let us agree to disregard the imaginary part.

For a Fourier series with fundamental period T, similar remarks apply and the general form becomes

$$\frac{1}{T} \sum_{r=0}^{\infty} C_r \exp\left(\frac{2\pi i}{T} rt\right). \tag{3.7}$$

>3.4 Sample values and periodicity

The objective is to find the coefficients C_r. In §3.1 it was speculated that if a periodic function $F(t)$ was known at N sample points, then it ought to be possible to represent it with exactly N harmonics; in other words we want to choose coefficients C_r so that

$$\frac{1}{T} \sum_{r=0}^{N-1} C_r \exp\left(\frac{2\pi i}{T} rt\right) \tag{3.8}$$

takes exactly the same values as $F(t)$ at sample t values. It is convenient to have a notation for these values. The interval between them is T/N; call this Δt. The sample T values are integer multiples of this,

$$t = s\Delta t$$

where s can be positive or negative. By taking $s = 0,1,2,...$ up to $\frac{1}{2}N - 1$ (suppose N is even), all the sample points in the positive range $0 < t < \frac{1}{2}T$ are obtained. By taking $s = -\frac{1}{2}N, -\frac{1}{2}N + 1,..., -1$, all the points in the negative range, $-\frac{1}{2}T < t < 0$ are obtained. Because of the periodicity of $F(t)$, these negative t values will be the same as the values obtained at $s = \frac{1}{2}N, \frac{1}{2}N + 1,..., N - 1$. The N sample values may therefore be written as

$$F_s = F(s\Delta t), \quad s = 0,1,2,..., N - 1. \tag{3.9}$$

Combining this with (3.8) our objective is to find coefficients C_r such that

$$F_s = \frac{1}{T} \sum_{r=0}^{N-1} C_r \exp\left(\frac{2\pi i}{T} r \, s\Delta t\right). \tag{3.10a}$$

Replacing Δt with its value T/N gives

$$F_s = \frac{1}{T} \sum_{r=0}^{N-1} C_r \exp\left(\frac{2\pi i}{N} rs\right). \tag{3.10b}$$

The periodicity of F is reproduced precisely by the exponential terms in (3.10b) and (3.8). Taking the latter first, adding T to t changes

$$\exp\left(\frac{2\pi i}{T} rt\right)$$

to

$$\exp\left(\frac{2\pi i}{T} r(t + T)\right)$$

which equals

$$\exp\left(\frac{2\pi i}{T} rt + 2\pi i \, r\right).$$

The extra $2\pi i r$ in the argument does not change the value, as may be seen from equation (3.3) by adding 2π, or any multiple of 2π, to A. By similar reasoning adding N to s in (3.10b) has no effect on the value.

The form of (3.10b) should provoke another observation: in the argument of the exponential, r and s appear on an equal footing, suggesting periodicity in r as well as s. This is indeed true, and is closely tied up with the interpretation of the coefficients C. Recall that C_r is the (complex) coefficient of the harmonic with frequency r times the fundamental; but it was shown that the highest frequency that it was sensible to take was $\frac{1}{2}$ of the sampling rate. The sampling rate equals N/T, the fundamental frequency is $1/T$, and the coefficient C_r is associated with the harmonic of frequency r/T (measuring all frequencies in cycles per unit time). So the highest sensible r value should be $\frac{1}{2}N$. However, the argument that led to this limit depends only on the magnitude of frequency, not its sign. Indeed, for the cosine Fourier series originally considered, the idea of negative frequency would be pointless because cosines are even functions and so if q is a frequency

$$\cos(2\pi q\, t) = \cos(2\pi(-q)t).$$

But now that complex harmonic functions are being considered, the idea of negative frequency has more point, because $\exp(2\pi i q t)$ is not equal to $\exp(2\pi i(-q)t)$ as the sines which make up the imaginary part of this exponential are odd functions. Therefore, if some coefficients of negative frequencies were introduced into the complex Fourier series (3.10b) or (3.8), they would carry different information from the coefficients of positive frequencies. This suggests that rather than (3.8) we should have tried

$$\frac{1}{T}\sum_{r=-\frac{1}{2}N}^{\frac{1}{2}N-1} C_r \exp\left(\frac{2\pi i}{T} rt\right) \tag{3.11}$$

and instead of (3.10b)

$$F_s = \frac{1}{T}\sum_{r=-\frac{1}{2}N}^{\frac{1}{2}N-1} C_r \exp\left(\frac{2\pi i}{N} rs\right). \tag{3.12}$$

The observation that (3.8), (3.10b), (3.11) and (3.12) are periodic in r is all that is necessary to reconcile the two sets of equations. Just as it was seen that the sample values $F_0, F_1, \ldots, F_{N-1}$ could be thought of as a part of an infinitely repeating periodic sequence, it can be imagined that the coefficients C_r are also part of a periodic sequence. The periodic nature of (3.12) implies that any consecutive set of N values can be used to evaluate the sum. However, we will only expect that frequencies in the range $-N/2T$ to $+N/2T$ (r values between $-\frac{1}{2}N$ and $+\frac{1}{2}N$) would bear any relation to actual Fourier series harmonics.

The choice of which set of values to use as standard is evidently arbitrary, and because most computer languages do not handle negative subscripts, we will choose to work with the standard range $r, s = 0, 1, 2, \ldots, N-1$.

For s values, values $s = 0,1,...,\frac{1}{2}N - 1$ correspond to values of t from 0 to $(\frac{1}{2}N - 1)\Delta t$, and we will choose to associate values $s = \frac{1}{2}N, \frac{1}{2}N + 1,...,N - 1$ with values of t from $-\frac{1}{2}T$ to $-\Delta t$. For r values, we will similarly associate $r = 0,1,...,\frac{1}{2}N - 1$ with frequencies $0, 1/T,...,(\frac{1}{2}N - 1)/T$, and $r = \frac{1}{2}N$, $\frac{1}{2}N + 1,...,N - 1$ with frequencies $-\frac{1}{2}N/T, (-\frac{1}{2}N + 1)/T,..., -1/T$.

With this convention in mind, look back at one of the displays produced by FOURIER, and this time take note of the real and imaginary plots on each graph. For the examples that you ran, imaginary parts were zero in both cases. In the top graph, this is true because we chose $F(t)$ to be a real function. Because it was also chosen to be even, only cosine harmonics were needed to build it up, and so the imaginary parts of the DFT (namely of all the coefficients C_r) were all zero (see (3.6)). Notice also that the DFT is an even function of frequency; given that all the b_r are zero, an even DFT is needed to ensure that (3.12) has a zero imaginary part when evaluated. This is because $\exp(-ix) + \exp(ix) = 2\cos(x)$; the imaginary parts of each pair at positive and negative frequencies cancel out, and the real parts double up. Recall that we have already met this factor of 2 in our earlier comparison of the DFT with a Fourier series.

$\rangle$3.5   Finding the discrete Fourier transform

It is time now to look at the answer to the question of how to find the DFT. Repeating (3.10b), it is required to find C_r so that

$$F_s = \frac{1}{T} \sum_{r=0}^{N-1} C_r \exp\left(\frac{2\pi i}{N} rs\right). \tag{3.13}$$

The answer has a symmetry which should appeal to the aesthetic sense of all mathematicians:

$$C_r = \frac{T}{N} \sum_{s=0}^{N-1} F_s \exp\left(-\frac{2\pi i}{N} sr\right). \tag{3.14}$$

These relationships are exact inverses of one another. That is, starting with any set of values F_s (N complex values), (3.14) can be used to obtain N complex values C_r; these could then be substituted in (3.13) and the original N complex values F_s would be obtained. There is a comparison with a coding (encrypting) machine; think of F_s as the message, (3.14) puts it in code, (3.13) unscrambles the code. The proof of this exact relation is left to an appendix.

Run program DFT, starting with $M = 3$, $T = 4$, to illustrate some of the properties that have just been discussed. DFT stores complex values in a 2-subscript array called F(I,J). J = 0 for real parts, J = 1 for imaginary parts. First DFT calculates N sample values of the function, (as supplied, it uses the function defined in (3.1)), setting all imaginary parts to zero. Then it prints the values, distinguishing between the lower half $I = 0,1,...,\frac{1}{2}N - 1$, and the upper half $I = \frac{1}{2}N,...,N - 1$. Next, it calculates the DFT; these calculated values replace the values previously held in F(I,J), and are printed in exactly the same way. Finally, the DFT is inverted, and once more the values are printed. You can see that to the accuracy of the computer, the original values are exactly recovered.

There are many interesting calculations to be done using FOURIER and DFT. First, finish off the investigation of the relationship of the DFT of $F(t)$ (defined in equation (3.1)) to the modified Fourier series (3.2b); recall that we were comparing $2/T$ times the DFT values with the terms of this series. Write a short program of your own, or use a calculator, to work out the first few terms of the series, then multiply them by $T/2$ and keep the results handy. Now run DFT and compare the values; keep $T = 4$ and try $M = 3$ upwards for as far as you have time for. Alternatively modify DFT so that instead of the imaginary part of C_r it prints $T/(\pi C_r)$; if the C_r were in agreement with prediction this expression would give the sequence 1, 0, $-3, 0, 5, 0, -7,...$. Either way you should be convinced that as the number of sample points increases, the agreement gets better and better. You can easily modify program DFT so that it works with other functions, for example those at the end of Part 2. Try some with odd parity, and some which are neither odd nor even. In every case notice how the inverse DFT is exactly equal to the original values to within very close tolerance, and observe how the symmetry of the original function affects results. You could also try generating a function with random sample values, and even that will inverse transform back to its original values.

The calculations that have been suggested are all designed to reinforce the idea that the infinite Fourier series is a useful concept, relevant to practical applications. In practice it is never possible to use an infinite number of points, and the discrete Fourier transform, which has properties similar to a finite Fourier series, is the technique that would be used in any practical calculation. But the infinite Fourier series represents a limiting form of the DFT; as you have seen, this form is approached, but never reached, so it can be thought of as the ideal that we would like to achieve if there was infinite time available; it becomes a standard against which we can compare any finite set of values obtained by using a DFT. Notice the manner in which this approach proceeds; except for the constant, all the harmonic coefficients change as the number of sample points increases, in contrast to the technique used in Part 2 where the coefficients were always the Fourier coefficients and the approach to the ideal was made by increasing the number of terms used in the calculation.

Another important feature to watch for in the computer runs is the relative rates at which the DFT coefficients approach the limiting values for different choices of function. The on–off function originally chosen makes this approach quite slowly; for example to obtain the coefficient at frequency 11 times fundamental, to better than 0.01, it is necessary to take $M = 7$, whereas for the triangular waveform similar accuracy is achieved by $M = 5$. Incidentally, accuracy here must be measured relative to the leading terms in the series, all of order unity in these examples.

⟩3.6   The Fourier series limit

Knowledge of the DFT formula makes it possible to calculate the values of the Fourier series coefficients, which up to now have simply been quoted. The necessary formulae can be obtained (without mathematical rigour, but very plausibly) using only elementary calculus and the idea of integration as the area under a curve.

Go back to equation (3.14), replace T/N by Δt, to obtain

$$C_r = \sum_{s=0}^{N-1} F(s\Delta t) \exp\left(-\frac{2\pi i}{T} r\, s\Delta t\right) \Delta t. \tag{3.15}$$

This sum can be envisaged as the total area of N vertical rectangles, each of width Δt, with corners at the t values $0, \Delta t, 2\Delta t, \ldots, (N-1)\Delta t$, and heights $F(t)\exp[-(2\pi i/T)rt]$. Under suitable conditions, as N is chosen larger and larger, this sum becomes closer to the value

$$\int_0^T F(t) \exp\left(-\frac{2\pi i}{T} rt\right) dt \tag{3.16}$$

from which the formulae for the Fourier coefficients follow

$$a_0 = \frac{1}{T} \int_0^T F(t)dt \tag{3.17}$$

$$a_r = \frac{2}{T} \int_0^T F(t) \cos\left(\frac{2\pi}{T} rt\right) dt \tag{3.18}$$

$$b_r = \frac{2}{T} \int_0^T F(t) \sin\left(\frac{2\pi}{T} rt\right) dt \tag{3.19}$$

where $r = 1, 2, \ldots$ and the a_r and b_r are the actual cosine and sine Fourier series coefficients (the doubling up has been done). Notice that a_0 is just the 'average value' of $F(t)$ over one period.

$\rangle$3.7 The half-way values at jumps

In equation (3.1) care was taken to define $f(t)$ at the 'jumps' at $t = +1$ and -1. These *spot values* will have no effect on the values of the integrals (3.17) to (3.19), and omitting them, or making them into any other value, will not affect the Fourier series coefficients. Why then do they need special care?

The answer is to do with the way that Fourier series handle jumps. In Part 2, we started by expecting that any function represented by a Fourier series would have to be smooth. It was then shown that by taking more harmonics it was possible to make jumps more and more rapidly until they were completed in a time interval which was smaller than the resolution required. Using a finer resolution would always reveal the continuous nature of the jump, and in particular any finite Fourier series will have to have a unique and well defined value at the position of the jump. It turns out that the limit of this value is always the average of the values on either side; in the example being used (3.1), these values are 0 and 1 so the limiting value of the Fourier series at $t = -1$ and $+1$ is expected to be $\frac{1}{2}$. The half-way value property is easy to verify for any of the examples quoted at the end of Part 2; for the function (3.1) we have seen that only odd harmonics occur, each having the form

$$a_{2n+1} \cos\left(\frac{2\pi}{T}(2n+1)t\right)$$

where $T = 4$, and therefore at $t = 1$ the argument of the cosine is an odd multiple of $\pi/2$, which means that the cosine takes the value zero. This is true for all terms except the constant, which was $\frac{1}{2}$, verifying the property in this case.

Although an isolated *peculiar* value in $f(t)$ will not affect the integrals (3.17) to (3.19) (nor will a finite number of such values), a change in a single member of the sample values F_s obviously will affect the DFT, so there appears to be an inconsistency. It is not hard to see that the effect of an isolated value will diminish as more and more sample values are taken, so the limit approached will turn out to be the same, and in this sense the isolated value does not matter. But for practical purposes we want to approach the limit as fast as possible, so that within whatever limits of time and memory the computer imposes, as accurate as possible a result can be obtained. It turns out that the best way to do this is to 'fool' the DFT algorithm by using the half-way values at jumps.

If you know about the rounding error problems that computers have with real numbers, you will known that there may be problems in detecting these spot values. For example, the obvious way to program a function like (3.1) is something like this:

```
IF  ABS(t) <1  THEN  func=1
IF  ABS(t) =1  THEN  func=0.5
IF  ABS(t) >1  THEN  func=0
```

Unless the value of t is *exactly* 1.0, the spot value may never get used. In all the examples suggested here, however, the imposed period T, as well as N, will be a power of 2, and the sampling interval Δt can be stored exactly by the computer in these circumstances. Otherwise, some kind of tolerance must be used, for example:

```
IF  ABS(1 - ABS(t)) <0.00001  THEN  func=0.5
```

It is quite instructive to see the effect of getting the jump values wrong. It is very easy to modify FOURIER and DFT to omit the value at $t = +1$ or -1, for example:

```
IF  ABS(t) < =1  THEN  func=1  ELSE  func=0
```

Try repeating some of the previous calculations with this change, and remember to try similar changes later in this part. One interpretation of

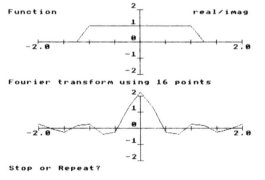

Figure 3.3 The effect of modifying FOURIER as suggested at the end of §3.7, so that $f(t)=1$ at $t = 1$. Compare with figure 3.2, where $f(t)=\frac{1}{2}$ at $t = 1$.

the results is that the DFT has been fooled into thinking that the jumps occur at $1 + \frac{1}{2}\Delta t$ and $-(1 + \frac{1}{2}\Delta t)$; as N gets larger, Δt gets smaller and the limit is approached as before. Now, however, the function does not spend equal times up and down, which as you should see affects results considerably.

>3.8 Fourier transforms

The way in which a discrete Fourier transform is related to a Fourier series has been demonstrated. In the computer runs, this was done by keeping T fixed and increasing the number of sample points, and it was discovered that although the number of values in the DFT increased without limit, any individual value gradually approached a limit.

Another kind of limiting behaviour will now be demonstrated. To give some purpose to this, recall the light switch example at the end of Part 2, which caused radio interference. We noted at the time that one unrealistic feature of our model was that we had to make the switch go on and off indefinitely, and at regular intervals. A more realistic model would be the function $f(t)$ defined in equation (3.1) (never mind about the values of $\frac{1}{2}$ at the switch-over time), because this is not periodic. Now everything connected with Fourier series and DFT has periodic behaviour built into it, so at first sight none of our theory will be any help here. But there is something we can do: we can alter the period that was forced into $f(t)$.

Earlier in this part, we worked entirely with T fixed at 4, which produced a *periodised* version of $f(t)$ which spent equal times in the 'on' and 'off' states. By choosing larger values of T, this balance can be changed. For example, choosing $T = 8$ generates a periodic function which is 'off' (value zero) from $t = -4$ to -1, $t = 1$ to 4, and 'on' (value 1) at the middle times $-1 < t < 1$, so that now it is only 'on' for 1/4 of the time. Clearly we can continue to extend the period in this way as much as we like, the hope being that by making it sufficiently large our periodised function will behave indistinguishably from the ideal model function (3.1).

Program FOURIER allows us to do this and observe the effects. Before running it however, let us try to predict what will happen. First, note that if the sampling rate is to be kept at a reasonable level, we must increase N as we increase T (because the sampling rate is N/T). The fundamental frequency is $1/T$, so increasing T implies that the fundamental is at a lower frequency, and as all the harmonic frequencies are multiples of this, the frequency interval between harmonics will be smaller. For example,

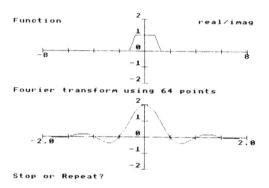

Figure 3.4   Results from FOURIER with $M = 6$, $T = 16$. See §3.8.

choosing $T = 8$ will give a frequency interval of 1/8 compared with 1/4
before, so our new set of harmonics will include the old set. The maximum
frequency is half the sampling rate, which we will keep constant by
increasing N. The frequency range will then also remain constant at N/T.

With this in mind, run FOURIER using the following sequence of
values:

$$M = 4 \quad T = 4$$
$$M = 5 \quad T = 8$$
$$M = 6 \quad T = 16$$

etc.

The results are surprising. The form of the lower (DFT) graph does not
change dramatically, but instead appears to become smoother whilst
continuing to go through all the values that it went through before. In
particular, notice that the value at zero frequency remains at 2. Now an
explanation can be given for the presence of the $1/T$ factor when we first
wrote down (3.6), subsequently affecting the equations (3.13) and (3.14). If
the $1/T$ was included in the coefficients C_r, they would all get smaller as T
got larger, and the graph of the DFT would appear to decrease to zero.
Instead, it now appears that the values C_r are actually sample values
drawn from some underlying function, just as the values F_s were drawn
from the originally defined function $f(t)$. Only a certain 'window' of values
of this function can be seen, of total width equal to the sampling frequency.
To see more of it, increase the sampling frequency, for example as in this
set of values:

$$M = 5 \quad T = 4$$
$$M = 6 \quad T = 8$$
$$M = 7 \quad T = 16.$$

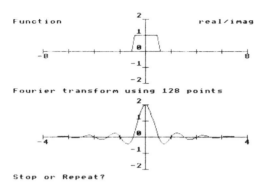

Figure 3.5   Results from FOURIER with $M = 7$, $T = 16$. See §3.8.

Repeat these two sets of calculations using program DFT. Start with $M = 4$, $T = 4$ and note the first eight DFT values carefully. Then repeat with $M = 5$, $T = 8$; compare values numbered 0,2,4, etc with your noted values and observe how well they match (there should only be slight variations in the last one or two decimal places). Continue with $M = 6$, $T = 16$, etc and note how well agreement continues. Now use $M = 5$, $T = 4$; this time you will find that the first eight values of the DFT are only approximately in agreement with your written down values, but if you make a new note of this set, and then continue to compute with $M = 6$, $T = 8$, etc subsequent values will mesh in with the new set.

In other words, the underlying function depends on the sampling rate. The search for a more absolute underlying function requires that both the sampling rate and T are increased, and it turns out that if this is done then a true limiting function can be found, and is known as the *Fourier transform* of $f(t)$.

The computer runs already done give a rough idea of what it looks like for our choice of $f(t)$. To get a better idea, run FOURIER with the following sequence of values, which doubles sampling frequency and T at each stage:

$$M = 3 \quad T = 4$$
$$M = 5 \quad T = 8$$
$$M = 7 \quad T = 16.$$

Notice that M goes up by 2 at each stage, quadrupling the number of points. Repeat with DFT to check the actual numerical values; compare values whose index numbers are given along each row of the following table.

M =	3	5	7	9	11	Frequency	Limiting
T =	4	8	16	32	64		value
Index no.	0	0	0	0	0	0	2
	1	2	4	8	16	1/4	1.273
	2	4	8	16	32	1/2	0
	3	6	12	24	48	1/3	−0.424

The final column gives the limiting values that you should find are approached. Each row corresponds to a fixed frequency, which is shown in the penultimate column.

⟩3.9   The Fourier transform as a limit

A general mathematical form for the Fourier transform can be found from equations (3.12) and (3.14) in a manner similar to that which we used to obtain Fourier series coefficients. The only difference is that this time in (3.14) T is not constant, so just as we wrote

$$F_s = f(t) \quad \text{where } t = s\Delta t$$

we now write

$$C_r = C(q) \quad \text{where } q = r\Delta q, \qquad \Delta q = 1/T$$

and note that

$$\frac{sr}{N} = \left(\frac{Nt}{T}\right)(qT)\frac{1}{N} = qt.$$

Then (3.14) can be written

$$C(q) = \sum f(t) \exp(-2\pi i q t)\Delta t \tag{3.20}$$

which has the limiting form

$$C(q) = \int_{-\infty}^{+\infty} f(t) \exp(-2\pi i q t)\, dt. \tag{3.21}$$

The limits of the integral are infinite because we are in effect integrating over one infinite period. Equation (3.13) can similarly be seen to be

$$f(t) = \int_{-\infty}^{+\infty} C(q) \exp(2\pi i q t) dq. \tag{3.22}$$

For the function $f(t)$ which has been used all along, the integral in equation (3.21) is easy to evaluate and gives

$$C(q) = \frac{\sin(2\pi q)}{\pi q}.$$

This is how the limiting values were calculated; for example

$$C(\tfrac{1}{4}) = \frac{\sin(\pi/2)}{\pi/4} = \frac{4}{\pi} \quad C(\tfrac{1}{2}) = \frac{\sin \pi}{\pi/2} = 0 \quad C(\tfrac{3}{4}) = \frac{\sin(3\pi/2)}{3\pi/4} = -\frac{4}{3\pi}.$$

Evaluating $C(0)$ requires a little care because there is a q in the denominator; it is necessary either to use a series expansion for $\sin(2\pi q)$ for small values of q, or to note from (3.21) that

$$C(0) = \int_{-\infty}^{+\infty} f(t)\, dt \tag{3.23}$$

which is trivially 2 in this case.

A note of caution should be sounded about the mathematical rigour of our approach here. In writing down (3.21) and (3.22), we have implicitly assumed that the process of increasing sampling rate and periodicity does indeed lead to a set of finite values for the DFT. For all functions suggested, this is the case, and it will also be true for almost every function that models a real physical situation correctly (infinite values do not occur in nature). However, even when considering cases where there is no finite limit, it is sometimes convenient to pretend that there is; the formulae (3.21) and (3.22) can be useful and meaningful in many such cases, which are dealt with in many standard texts.

$\rangle$3.10   Fourier transforms of a periodic function

One such case will be considered here because it illustrates both the point given above and the relation between Fourier transforms and Fourier series. Suppose we try to take the Fourier transform of a periodic function — what will happen? There is plenty of room here for confusion, because a non-periodic $f(t)$ has a periodised function $F(t)$ associated with it. The difference is that $F(t)$ will change as the imposed period T increases, whereas if $f(t)$ is itself periodic, the periodised function $F(t)$ will not change. The effect can easily be illustrated by some simple changes to program

FOURIER. Change the function definition so that

$$f(t) = \begin{cases} 0.5 & \text{if } t = 2n + 1 \quad \text{(an odd integer)} \\ 1 & \text{if } 4n - 1 < t < 4n + 1 \\ 0 & \text{if } 4n + 1 < t < 4n + 3. \end{cases} \qquad (3.24)$$

Also change the span of the vertical axis so that it goes from $-T/2$ to $+T/2$, and repeat the last sequence of runs (i.e. $M = 3$, $T = 4$, then $M = 5$, $T = 8$, etc) (program FOURIER2 has the necessary changes incorporated). The first run should give results exactly the same as with the original version, because the periodised function $F(t)$ is the same in both cases. But as T is increased, $F(t)$ becomes different and the transform changes. Instead of the development of a smoother function, as we observed before, you should see the development of a spikey function. The height of the spikes will grow with T (but the graphics routines will use new scales if the program modifications have been made as suggested). The width of the spikes decreases, even though the sampling frequency is increasing, and it does not appear that any limit is being approached. Repeat with similar changes to DFT (or use DFT2). Now you can see the values, and it is clear that the spikey nature of the graph is not just approximate; it is as exact as computer precision allows. Work out the frequencies at which the spikes occur; you should find that they are all multiples of the fundamental frequency of the function itself (as opposed to the fundamental frequency of the periodised function). What is more, the values should be $T/4$ times the values that were obtained when we first used the DFT to obtain a Fourier series of the periodic function (3.1).

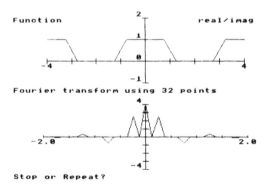

Figure 3.6 Results from FOURIER2 with $M = 5$, $T = 8$, showing the spiky nature of the DFT of a periodic function. See §3.10.

It is possible to see why by looking at (3.14). Compare the calculations (for example) with $M=4$, $T=4$ using either (3.1) or (3.24) (it makes no difference as $T=4$), and $M=5$, $T=8$ using (3.24). Take $r=0$ to start with; from (3.14)

$$C_0 = \frac{8}{32} \sum_{s=0}^{31} F_s \exp(0).$$

Because the sampling frequency has been kept the same, whilst T was doubled, F_s just goes through two cycles of the values that it takes when $M=4$, and as the ratio T/N is unchanged, C_0 is twice what it was. For C_{2r} (i.e. even numbered C), write

$$C_{2r} = \frac{8}{32} \sum_{s=0}^{31} F_s \exp\left(-\frac{2\pi i}{32} s\, 2r\right)$$

$$= \frac{8}{32} \sum_{s=0}^{31} F_s \exp\left(-\frac{2\pi i}{16} rs\right).$$

Because a factor of 2 could be cancelled in the argument of the exponential, the expression can be reduced to the same as that for the $M=4$, $T=4$ case except that the sum is over twice as many values. The F_s are known to go through two cycles as s runs from 0 to 31, and on inspection, it is easily seen that the values of $\exp[-(2\pi i/16)rs]$ do as well. So once again, (new C)=(old C) times (number of actual periods in the imposed period T).

It is also possible to prove that the odd numbered C are zero, but that is left as an exercise for the more adventurous. It is intuitively obvious that it must be so, because a periodic function should be made up only of harmonics which have whole number multiple frequencies of the fundamental. By the same argument, if T was chosen to be 16, we would only expect every fourth coefficient to be non-zero (and of course some of those could be zero also).

⟩3.11 Discussion

You can now perhaps appreciate that a Fourier transform is a more general concept than a Fourier series. The former effectively tells us how to build up an arbitrary function by adding together a large number of harmonics which are only infinitesimally spaced out in frequency. The spacing is $1/T$, which gets smaller and smaller as the imposed period T gets larger. The amount of a particular harmonic to be added in is given by C_r/T, or C_r times the spacing. As T increases, neighbouring harmonics are included which keeps the total amount at the right strength; the contributions are spread out over a finite and continuous frequency range. For a periodic function, the contributions have to be concentrated at exactly the periodic frequency and multiples of it, and to compensate they must get higher. The amount of a contribution can conveniently be measured by area; look again at the series of spikes generated by FOURIER2 and notice that for each triangular spike, its height is proportional to T, and its base to $1/T$, and therefore its area remains constant as T gets larger. The Fourier series coefficients are the areas under the corresponding spike (after allowing for doubling up due to symmetry as before).

A good physical illustration of a Fourier transform is a spectrum. Light from a source such as a heated filament lamp bulb can be broken up into its constituent frequencies (colours) and will produce a smeared out spectrum. The physicist will measure it in terms such as energy flow between frequency a and frequency b, for a whole range of adjacent frequencies. On the other hand, light from a fluorescent source, such as a sodium lamp, will produce a spectrum with a finite number of very distinct lines, corresponding to the statement that all the energy is concentrated at a few frequencies. If the spectral lines were of zero width they would have to have an infinite local strength; in practice, they are more like bands. Of course no light source can be truly periodic, as it would have to persist for an infinite length of time; the lack of true periodicity shows up in the fact that spectral bands always have a finite, non-zero width, and there are other physical reasons why these should be *spread*, such as the motion of the fluorescing atoms themselves.

To return to the radio interference example, the turning on of a switch generates a whole spectrum of radio waves, rather than the discrete harmonics that appeared before. The practical effect however is much the same because radio detector circuits actually respond to a band of frequencies rather than just one exact frequency. Our earlier simplification was justified because the spectrum decays proportionally to 1/(frequency), just as the Fourier series coefficients of the series.

⟩3.12 Some further examples

All the following examples have finite Fourier transforms; most can be calculated analytically. Modify FOURIER3 to calculate them, using a

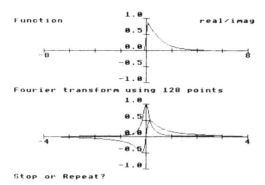

Figure 3.7   Results from FOURIER3 for example (ii), §3.12.

progression of M and T values until you have a good idea of the final shape (suitable values are suggested, but try others). FOURIER3 is a slightly modified version of FOURIER which allows the range of the y axis to be changed by data input at run-time, which is more convenient than using a fixed scale.

(i) Even exponential decay

$$y=\exp(-|t|) \quad C(q)=2/(1+4\pi^2q^2).$$

Try $(M,T)=(3,4)$, $(5,8)$, $(7,16)$,....

(ii) Switch-on, exponential decay

$$y=\begin{cases} 0 & t<0 \\ \frac{1}{2} & t=0 \\ \exp(-t) & t>0 \end{cases} \quad C(q)=1/(1+2\pi iq)$$

Try $(M,T)=(3,4)$, $(5,8)$, $(7,16)$,....

(iii) Switch-on, faster exponential decay

$$y=\begin{cases} 0 & t<0 \\ \frac{1}{2} & t=0 \\ \exp(-2t) & t>0 \end{cases} \quad C(q)=1/(2+2\pi iq)$$

Try $(M,T)=(3,4)$, $(5,8)$, $(7,16)$,....

Compare the last two examples, noting how scale is affected.

(iv)   Switch-on, hyperbolic decay

$$y = \begin{cases} 0 & t<0 \\ \tfrac{1}{2} & t=0 \\ 1/(1+t) & t>0 \end{cases} \quad \text{no simple analytic form}$$

Try $(M,T) = (3,4)$, $(5,8)$, $(7,16)$,....

(v)   Transform of 'top hat' function

$$y = \frac{\sin(2\pi t)}{2\pi t} \quad C(q) = \begin{cases} 0 & |q|>1 \\ \tfrac{1}{4} & |q|=1 \\ \tfrac{1}{2} & |q|<1 \end{cases}$$

Try $(M,T) = (3,2)$, $(5,4)$, $(7,8)$,....

This example illustrates the property that if $C(q)$ is the transform of $y(t)$, then $y(p)$ is the transform of $C(t)$. (Note that the use of p and t as function arguments is just a convention.) $y(t)$ above is the theoretical transform of the 'top hat' function which was used in many of the illustrated examples. See for instance figure 3.5.

(vi)   Triangle function

$$y = \begin{cases} 0 & |t|>1 \\ 1-|t| & |t| \leqslant 1 \end{cases} \quad C(q) = [1 - \cos(2\pi q)]/8\pi^2 q^2.$$

Try $(M,T) = (3,4)$, $(5,8)$, $(7,16)$,....

Compare $C(q)$ here with $y(t)$ in the previous example. Note that $C(q)$ decays much more rapidly as q increases than $y(t)$ does as t increases. This is because the 'top hat' function is discontinuous, whereas the triangle function is continuous. See also §2.7, where similar properties are noted for Fourier series coefficients.

(vii)   Gaussians

$$y = k \exp[-(kt)^2] \quad C(q) = \sqrt{\pi} \exp[-(\pi q/k)^2].$$

First use $k=2$ and try

$$(M,T) = (3,4), (5,8), (7,16),...$$

then repeat with $k = 2,4,...$ and as high as you can. Notice how the width of the humps is related. Notice that a Gaussian transforms to another Gaussian.

(viii) Shifted Gaussian

$$y = 2 \exp[-4(t-v)^2] \quad \text{with } v = 1.$$

Use $(M, T) = (5,8), (7,16), \ldots$

then repeat with $v = 2$.

The $(t-v)$ term has the effect of shifting the original function. Try 'shifting' some of the other examples. How does a shift in t space affect the Fourier transform?

Many other analytical examples are set as exercises in text books.

> Appendix 1

> The FFT Algorithm

The discrete Fourier transform (DFT) and its inverse (3.13) and (3.14) are of similar form, and apart from the overall constants can both be evaluated by a single procedure. The form is

$$G_s = \sum_{r=0}^{N-1} F_r \exp\left(-\frac{2\pi i}{N} rs\right) \quad s = 0, 1, \ldots, N-1 \tag{A.1}$$

where the only change needed to obtain the inverse form is to change the sign of the argument of the exponential function.

In theory this sum is defined for any value of N but a particularly efficient algorithm for evaluating it on a computer exists when N is a power of 2. The algorithm is known as the *fast Fourier transform* algorithm (FFT).

Since there are N values of G_s to be calculated, and each value involves evaluating and summing N terms, the obvious technique for programming the DFT will involve approximately N^2 such evaluations. By contrast, the FFT needs approximately $\frac{1}{2}N\log_2 N$ evaluations. For values of N below 64 (say), the difference hardly matters, but for higher values it becomes very significant.

Here is an outline of how the algorithm works.

First note that equation (A.1) does not involve T; in other words the values G_s are defined by (A.1) without needing to know that the values F_r are sample values of a function. The set of values F_r, plus a value for N, are sufficient to define the G_s.

Now divide the values F_r into two groups, even numbered and odd numbered. Let

$$F'_j = F_{2j} \qquad F''_j = F_{2j+1}$$

where j runs from 0 to $\frac{1}{2}N - 1$.

Equation (A.1) can now be applied separately to each set of values F'_j, F''_j, using a value of N equal to half its original value. This will yield two sets of values G'_k, G''_k, where k also runs from 0 to $\frac{1}{2}N - 1$.

It is not hard to show that

$$G_k = G'_k + \exp\left(-\frac{2\pi i}{N}k\right)G''_k \qquad G_{k+\frac{1}{2}N} = G'_k - \exp\left(\frac{2\pi i}{N}k\right)G''_k. \quad \text{(A.2)}$$

A similar result holds for the inverse form, with the sign of the exponent changed.

Equation (A.2) tells us how to carry out *successive doubling*: from the DFT of $\frac{1}{2}N$ values we can obtain the DFT of N values. This process can be continued, each group F'_j, F''_j being repeatedly subdivided until N groups are obtained, each containing one value. Now equation (A.1) becomes completely trivial for $N = 1$, giving just

$$G_0 = F_0. \quad \text{(A.3)}$$

Now it is possible to work backwards, obtaining the DFT of successively double the group size.

As an example, consider how the set of eight values $v_0, v_1, v_2, v_3, v_4, v_5, v_6, v_7$ would be subdivided.

First decomposition, yielding two groups of four values each

$$(v_0, v_2, v_4, v_6), \quad (v_1, v_3, v_5, v_7).$$

Second decomposition, yielding four groups of two values each

$$((v_0, v_4), (v_2, v_6)), \quad ((v_1, v_5), (v_3, v_7)).$$

Third decomposition, yielding eight groups of one value each

$$(((v_0), (v_4)), ((v_2), (v_6))), \quad (((v_1), (v_5)), ((v_3), (v_7))).$$

The algorithm starts by rearranging values in the decomposed order. In the example, the next step would be to compute the DFT of each of the four pairs at the second level, using equation (A.2) four times over with $N = 2$. From this, the DFT of each of the two groups of four at the first level can be found, and finally these are doubled up to give the final result. Clearly this process will work when N is any positive integer power of 2.

> Appendix 2

> Proof of the DFT Inverse Relations

(See §3.6) It is required to prove that if

$$C_r = \frac{T}{N} \sum_{s=0}^{N-1} F_s \exp\left(-\frac{2\pi i}{N} sr\right) \qquad (A.4)$$

defines a set of N values $C_0, C_1, \ldots, C_{N-1}$ from N given values $F_0, F_1, \ldots, F_{N-1}$, then $F_k = \tilde{F}_k$, where

$$\tilde{F}_k = \frac{1}{T} \sum_{r=0}^{N-1} C_r \exp\left(\frac{2\pi i}{N} kr\right) \qquad (A.5)$$

for $k = 0, 1, \ldots, N-1$.

The proof starts in the obvious way by substituting the values C_r into the right-hand side of (A.5), giving

$$\tilde{F}_k = \frac{1}{N} \sum_{s=0}^{N-1} F_s \left[\sum_{r=0}^{N-1} \exp\left(\frac{2\pi i}{N} r(k-s)\right) \right]. \qquad (A.6)$$

Suppose it holds that

$$\frac{1}{N} \sum_{r=0}^{N-1} \exp\left(\frac{2\pi i}{N} rj\right) = \begin{cases} 1 & \text{if } j = mN \\ 0 & \text{otherwise} \end{cases} \qquad (A.7)$$

where m is an integer. On the right-hand side of (A.6) write $j = (k-s)$, then it will be seen that as s runs from 0 to $N-1$, only the term for which $j=0$ (when $s=k$) survives, and so $\tilde{F}_k = F_k$ as required.

To prove (A.7), consider first $m=0$, so $j=0$. The result is obviously 1 as each term on the left of (A.7) is $1/N$, and there are N terms. The same thing happens when m is any other integer.

Now consider the case when j is not a multiple of N. Then j/N is a rational fraction, and j and N must either have no common factor, or else it is possible to find two non-zero integers p,q such that

$$j/N = p/q \tag{A.8}$$

and p,q have no common factor, and q is not unity (otherwise $j=pN$). There will also be an integer n such that $j=np$, $N=nq$. n will be 1 if j and N themselves have no common factor.

Now the left-hand side of (A.7) reduces to

$$\frac{1}{N}\sum_{r=0}^{N-1}\exp\left(\frac{2\pi i}{q}rp\right) = \frac{n}{N}\sum_{r=0}^{q-1}\exp\left(\frac{2\pi i}{q}rp\right) \tag{A.9}$$

because of the cyclic nature of the exponential. Now let

$$w = \exp\left(\frac{2\pi i}{q}p\right).$$

We have $w^q = 1$, and furthermore, q is the smallest power for which this is true because p and q have no common factor.

Now $w^q = 1$ has q distinct complex roots, and they are w^r, $r=0,1,\ldots,q-1$. From (A.9), equation (A.7) can now be reduced to

$$\frac{n}{N}\sum_{r=0}^{q-1}w^r$$

which is zero because of the well known result that the sum of the roots of unity at any integer power is zero. This concludes the proof.

〉 Appendix 3

〉 Program Outlines

These program outlines give the main steps of each program in a descriptive manner which is not the same as a detailed program. There may be some variations in actual implementation, but these outlines will be a useful aid to understanding the actual programs.

Users wishing to adapt the programs are advised first to obtain a listing. In the case of BBC programs, note that the following library routines will be included in the supplied version and need not be listed as no modifications to them will be necessary:

> lines 9000–9999: the FFT procedure
> lines 10000 on : graphics routines.

〉Program SAMPLE1

Line numbers refer to the BBC BASIC program.

    Clear the screen

20 PRINT: "Program SAMPLE1"
           "Illustrates how a function may be"
           "represented by sampling its values,"
           "and the relation between this and"
           "screen resolution."

----
100   Define FNY(X), the function to plot
      DEF FNY(X) = X↑2

1010   Initial data:
        screenwidth : = 1024
        Initialise graphics procedures
          [note default x,y, ranges are − 10 to + 10]
        Set YHIGH : = 100 [for plotting y = x↑2]

2000   PRINT "Input a value for:"
                "Resolution"
        READ Resol
        check that Resol has a sensible value

2010   Clear screen, prepare for graphics
        Call procedure to set axes parameters

2020   PRINT "Resolution = ";Resol

2100   The chosen resolution is simulated by plotting squares.
        Calculate:
        Square size in graphics units SS : = screenwidth/Resol
        Square size in problem x-units SX : = SS/x-scale factor
        Square size in problem y-units SY : = SS/y-scale factor

2130   Now begin the square plotting process by starting with the
        outermost pair:
        Separation between squares dX = XHIGH − XLOW
        Plot square at (XLOW, FNY(XLOW))
        Plot square at (XHIGH, FNY(XHIGH))

2160   Now loop, filling in the next lot of squares at half the previous
        separation:

        REPEAT
            dX = dX/2: X = XLOW + dX

            REPEAT
                Plot square at (X,FNY(X))
                X = X + dX*2
                UNTIL X > XHIGH

PRINT "press SPACE to go on"
wait for SPACE to be pressed

UNTIL dX (= separation) less than (square width)/4

PRINT "Complete"

2270 END OF PROGRAM

3000 Define a procedure
      to draw square at (X,Y) using given resol.

      Round X,Y to be integer multiples of SX, SY:
      X : = SX*INT(.5 + X/SX)
      Y : = SY*INT(.5 + Y/SY)

      IF square size (= SS) is small (<5 say)
        THEN plot just one screen point (X,Y);
        ELSE fill in a square centred on (X,Y) with sides (SX,SY)

      END OF PROCEDURE

〉Program SAMPLE2

Line numbers refer to the BBC BASIC program.

      Clear the screen

  20 PRINT   "Program SAMPLE2"
          "Illustrates the effect of plotting"
          "a function using lines to join up"
          "the sample points."

          ----
          100   Define FNY(X), the function to plot
                DEF FNY(X) = X↑2

1010   Initial data:
       screenwidth : = 1024
       Initialise graphics procedures
          [note default x,y ranges are − 10 to + 10]
       Set max Y : = 100 [for plotting y = x↑2]

1900   Set the program phase indicator
       a$: = "B" (Beginning)

2000   PRINT "No. of sample points"
       READ N__samples
       check that N__samples is in range

2100   Clear screen, prepare for graphics
       Call procedure to set axes parameters

2110   PRINT N__samples "sample points."

2115   IF program phase a$ = "M" skip line plotting

                       Line plotting of the function:
           2120   Call PROClines
                  PRINT "continue(C) or repeat (R)?"
                  Call FNopt to see which option
                  chosen:
                  IF C then continue ELSE go to 2000

2500   Clear screen, prepare for graphics
       Call procedure to set axes parameters

2510   PRINT N__samples "sample points."

       Plot the function at max resolution as a series of points:
2520   Call PROCpoints
       then superimpose the line plot:
       Call PROClines
       PRINT "continue(C) or stop(S)?"
       Call FNopt to see which option chosen:
       IF S then clear screen and stop

ELSE PRINT "repeat to beginning(B) or middle(M)?"
Call FNopt and set a$ to "B" or "M"
Go back to 2000

3000   PROClines:
a procedure to plot FNY(X) by sampling at N__samples point
and joining up with lines

Set graph colour
Move to XLOW, FNY(XLOW)
Set sampling interval so that it is a whole number of graphics
units:
dX: = INT(XFA*(XHIGH − XLOW)/(N__samples − 1) + .5)/XFA
   where XFA is the x scale factor.

The main plotting loop:
X : = XLOW
REPEAT
    Draw to (X, FNY(X))
    X = X + dX
    UNTIL X > XHIGH + .5*dX

END OF PROCEDURE

3100   PROCpoints:
a procedure to plot FNY(X) by sampling it at points separated at
   screen resolution.

Set graph colour
Set sampling separation:
dX : = INT(.5 + screenwidth/xres)/XFA
   where XFA is the x-axis scale factor.

The main loop:
X = XLOW
REPEAT
    Plot a point at X,FNY(X)
    X = X + dX
    UNTIL X > XHIGH + .5*dX

END OF PROCEDURE

3200   FNopt(a$)
       A function to look at input from the keyboard until a character is
       typed which is in a$; then FNopt returns the position in a$.

       END OF FUNCTION

>Programs FSCOS, FSSIN, FSBOTH

Line numbers refer to the BBC BASIC version.
Flow is shown for FSCOS with minor modifications for FSSIN, FSBOTH

100   Dimension arrays:
      C(32)   for series coefficients
      C$(32)   for default values (string format)
      Y(160)   for sum of series sample values
      [add S( ) and S$( ) for FSSIN, FSBOTH]

110   Set default values and subscript limits

210   Clear the screen

220   PRINT   "COSINE HARMONICS"
              "Const"
      INPUT value for C   [not for FSSIN]
              "Up to what frequency"
              " multiple of 2.PI"
      INPUT value for N
              check N in sensible range

250   FOR I = 0 TO N − 1
              PRINT harmonic number and its default coefficient
              INPUT a new value or use default: C(I) = FNI(C$(I))
              [similarly for FSSIN, FSBOTH]
              NEXT I

300   PRINT   "How many periods to plot"
      INPUT value for NP and check
      PRINT   "Sampling rate"
      INPUT value for SF and check

500   Evaluation of the Fourier series:
      PRINT "Calculating"
      Set range shift rs:
          rs: = 0 if plotting even no. periods
          rs: = 1/2 if an odd number.
          Y0: = 0, Y1: = 0

      FOR I = 0 TO SF
          t = I/SF − rs
              (so range is 0 to 1 or −1/2 to +1/2)
          Y(I) = FNfs(t):   PRINT"."
          Keep running tally min and max values in Y0 and Y1
          NEXT

1000 Initialise for Graphics
      X-axis range is − NP/2 to + NP/2
      Y-axis range is Y0 to Y1
      Set scale factors and draw axes
      PRINT title including values of N and SF:
          "Cosines: max freq = " N "sampling = " SF
          [similar for FSSIN, FSBOTH]

2000 Do the plotting
      Set first colour or plotting style
      Set T interval dT: = 1/SF
      Move to first point at T = − NP/2, y = Y(0)
      FOR P = 1 TO NP (for each period)
          Flip the colour or plotting style
          FOR I = 1 TO SF: T = T + dT
              Draw to (T,Y(I))
              NEXT
          NEXT
      PRINT "press S to stop, C to continue"
      Call FNopt to see which option chosen:
      If S, stop; if C, go back to 210.

      END OF MAIN PROGRAM

3000 FNfs(t)
      a function to evaluate the Fourier Series

```
          LOCAL i (index), s (sum)
          s = C (the constant in cosine series)

          FOR i=1 TO N
              IF C(i−1) < >0 THEN s=s+C(i−1)*COS(2*PI*i*t)
              [modified for FSSIN, FSBOTH]
              NEXT
          return s as value of function.

          END OF FUNCTION
```

4000   FNI(V$)
       to input an expression or use default V$

```
          PRINT "(V$)"
          INPUT a string A$
          If A$ is null A$: = V$ (the default)
          Evaluate A$ and return its value

          END OF FUNCTION
```

4100   FNopt(a$)
       A function to look at input from the keyboard until a character is
       typed which is in a$; then FNopt returns the position in a$.

```
          END OF FUNCTION
```

〉Program FOURIER

Line numbers refer to the BBC BASIC program.
Program flow is shown for FOURIER with modifications for FOURIER2
and FOURIER3.

```
          Clear the screen

          Dimension the Fourier array:
          DIM F(127,1) for example.
```

```
  70   PRINT "FOURIER TRANSFORM"
       INPUT "Power of 2",M
           Check M in range
       INPUT "Range of T",T
           Check T positive
       [In FOURIER3, also input plotting ranges]

 100   N = 2↑M (no. of points: = 2 to power M)

 110   Clear screen, prepare for graphics
       Divide screen into an upper and lower sector
       Select upper sector
       Set Y-axis range from −2 to +2

 170   Call PROCsetF(M,T) to set values in array F.
       Call PROCgraphF(M,T) to plot these
       Print titles etc.

 260   Call PROCFFT(M,1) to work out the DFT:
           the result is in array F.
       Call PROCkF(M,T/N) to multiply values by T/N

 290   Select lower screen sector
       Call PROCgraphF(M,N/T) to plot the DFT
       Print title: "Fourier Transform using N points"

 330   At foot of screen print "Stop or Repeat?"
       Call FNopt to select option
       Then stop or repeat to 70

       END OF MAIN PROGRAM
----------------------------------------------------------------------------
1000   Function to define x(t)
       [This is modified for FOURIER2]
       May easily be changed by user.

       Standard function as described in text:
       DEF FNx(t)
       IF ABS(t) = 1 THEN = .5
       IF ABS(t) < 1 THEN = 1 ELSE = 0
----------------------------------------------------------------------------
```

2000   PROCsetF(M,T)
       procedure to set values in F(i,j).

       LOCAL n,n1,i,dt(n,n1 and i are integer)
       n = 2↑M: n1 = n DIV 2
       dt = T/n the t-sampling interval.

       Top half of array from t = − T/2 + dt to − dt:
       FOR i = n1 + 1 TO n − 1
           F(i,0) = FNx(dt*(i − n))
           F(i,1) = 0
           NEXT

       Bottom half of array from t = 0 to T/2:
       FOR i = 0 TO n1 − 1
           F(i,0) = FNx(dt*i)
           F(i,1) = 0
           NEXT

       Adjustment for any discontinuity:
       F(n1,0) = (FNx(dt*n1) + FNx(− dt*n1))/2
       F(n1,1) = 0

       END OF PROCEDURE

2190   PROCkF(m,k)
       procedure to multiply whole array by constant k

       LOCAL i
       FOR i = 0 TO 2↑m − 1
           F(i,0) = k*F(i,0)
           F(i,1) = k*F(i,1)
           NEXT

       END OF PROCEDURE

2270   PROCgraphF(m,T)
       procedure to plot graph of values in F(i,j)

       LOCAL i,j,n,n1,dt

n = 2↑m: n1 = n DIV 2: dt = T/n

Set x-axis range from −T/2 to +T/2
Call procedure to set axes parameters

Set style or colour for imaginary part (j=1)
  and plot.
Similarly for real part (j=0)

EXIT

To plot real or imaginary part:
MOVE to (−T/2,F(n1,j))
Plot values for t = −T/2 to −dt:
FOR i = n1 + 1 TO n − 1
    DRAW to ((i − n)*dt,F(i,j))
    NEXT
Plot values for t = 0 to T/2:
FOR i = 0 TO n1
    DRAW to (i*dt,F(i,j))
    NEXT

END OF PROCEDURE

4100   FNopt(a$)
       A function to look at input from the keyboard until a character is
       typed which is in a$; then FNopt returns the position in a$.

       END OF FUNCTION

⟩**Program DFT**

Line numbers refer to the BBC BASIC program.
Program flow is shown for DFT with modifications for DFT2 and DFT3.

Clear the screen

Dimension the Fourier array:
DIM F(2047,1) for example.

 70   PRINT "DISCRETE FOURIER TRANSFORM"
      INPUT "Power of 2",M
         Check M in range
      INPUT "Range of T",T
         Check T positive

100   N = 2↑M (no. of points: = 2 to power M)
      Set line count L to zero

170   Call PROCsetF(M,T) to set values in array F.
      Print title: "Original F(t)"
      Call PROCprintF(M) to print these

260   Call PROCFFT(M,1) to work out the DFT:
      the result is in array F.
      Call PROCkF(M,T/N) to multiply values by T/N

280   PRINT "Transform times T/N"
      Call PROCprintF(M) to print DFT

320   Call PROCFFT(M, −1) to work out the inverse DFT:
      the result is in array F.
      Call PROCkF(M,1/T) to multiply values by 1/T

330   PRINT "Inverse transform"
      Call PROCprintF(M) to print the inverse

400   PRINT "Stop or Repeat?"
      Call FNopt to select option
      Then stop or repeat to 70

      END OF MAIN PROGRAM

1000 Function to define x(t)
      [This is modified for DFT2]
      May easily be changed by user.

      Standard function as described in text:
      DEF FNx(t)
      IF ABS(t) = 1 THEN = .5
      IF ABS(t) < 1 THEN = 1 ELSE = 0

2000   PROCsetF(M,T)
       procedure to set values in F(i,j).

       LOCAL n,n1,i,dt (n,n1 and i are integer)
       n = 2↑M: n1 = n DIV 2
       dt = T/n the t-sampling interval.

       Top half of array from t = − T/2 + dt to − dt:
       FOR i = n1 + 1 TO n − 1
           F(i,0) = FNx(dt*(i − n))
           F(i,1) = 0
           NEXT

       Bottom half of array from t = 0 to T/2:
       FOR i = 0 TO n1 − 1
           F(i,0) = FNx(dt*i)
           F(i,1) = 0
           NEXT

       Adjustment for any discontinuity:
       F(n1,0) = (FNx(dt*n1) + FNx(− dt*n1))/2
       F(n1,1) = 0

       END OF PROCEDURE

2190   PROCkF(m,k)
       procedure to multiply whole array by constant k

       LOCAL i
       FOR i = 0 TO 2↑m − 1
           F(i,0) = k*F(i,0)
           F(i,1) = k*F(i,1)
           NEXT

       END OF PROCEDURE

3000   PROCprintF(m)
       procedure to print real & imag parts of F

```
      LOCAL I,N,N1
      N = 2↑m; N1 = N DIV 2

      PRINT "First half: 0 − (N1 − 1)"
      print column headings
      FOR I = 0 TO N1 − 1
          print F(I,0) and F(I,1)
          pause if linecount > 15 or half-range done
          NEXT

      PRINT "Second half: N1 − (N − 1)"
      print column headings
      FOR I = N1 TO N − 1
          print F(I,0) and F(I,1)
          pause if linecount > 15 or half-range done
          NEXT

      END OF PROCEDURE
```

4100   FNopt(a$)
       A function to look at input from the keyboard until a character is
       typed which is in a$; then FNopt returns the position in a$.

       END OF FUNCTION

> Program User's Notes

> Apple PASCAL Version

The purpose of these programs is fully described in the main text. This guide gives details of operations that are dependent only on the computing and programming aspects.

The programs are

SAMPLE1	FSCOS	FOURIER	FOURIER2	FOURIER3
SAMPLE2	FSSIN	DFT	DFT2	
	FSBOTH			

In addition, FFTUNIT (see below) holds various service routines.

All are written Apple PASCAL version 1.1. Operating familiarity with the Apple II computer and PASCAL system is assumed. The programs are stored on a disc named "FOURIER:", suitable for use with a two-disc drive system, and are held in both source (.TEXT) and compiled and linked (.CODE) form. A colour monitor will enhance the impact and educational value of the displays but is not essential.

>To execute the precompiled versions

Boot up the system with the "FOURIER:" disc on the second drive and use the X (execute) command. In response to the system prompt:

    EXECUTE WHAT FILE?

type (for example) "FOURIER:SAMPLE1".

>Text and graphics

For clarity of display and in order to let the user continue to see as much dialogue as possible, most of the dialogue will appear on the text screen, which is selected initially by all the programs. At appropriate intervals during the program, which will vary from program to program, there will be a pause in execution which is announced by the word PAUSE on both text and graphics screens. At such a pause, the user may select either the text screen (T), or the graphics screen (G), or elect to continue by pressing the space bar. Any keys other than these are ignored, and (T,G) may be used repeatedly as often as desired to switch between screens.

At any point, the program may force the selection of text or graphics if that is appropriate.

⟩Input of data

All data input is done using internal routines which check format; this is to prevent obscure program halts, but should any occur, the system may have to be re-booted and the program executed from the beginning again.

Integers must be input without leading spaces and must contain no internal spaces or characters other than digits, except that one + or − sign may be the first character. Any other characters will cause the message BAD INPUT — TRY AGAIN to appear, after which another attempt may be made. Integers may be terminated by space or return.

Real numbers may be input in any of three formats: integer, decimal, or rational fraction. Floating point form is not accepted. Decimal form is signed or unsigned, and has just one '.' which may appear immediately after the sign if desired; for example

$$+.5 \qquad -1.33 \qquad 0.1234 \qquad 3.141592.$$

Rational fractional form may be signed or unsigned, and has just one '/' appearing, which must be preceded by a valid integer with no intervening spaces; for example, $1/2$, $-1/5$, $22/7$.

Termination of input, and rules about spaces, are exactly as for integers. Violation of any of these rules leads to the BAD INPUT message and the chance to try again.

Data is often requested using *verify mode*. This prints the current value of the data item concerned before waiting for input. If the user then types a space or return, not preceded by any other character, the current value is left unchanged; this is a useful facility when the user wishes to make several program runs varying only a few of the data items. Otherwise a value is expected in one of the formats described above. Examples:

SAMPLING FREQUENCY   (*If user just types space or return,
(20)?   sampling frequency will remain set at
                              20*)

CONSTANT (-0.7854)? $-2/3$ (*will change value to $-2/3$*)

No distinction is made between occasions when integers or real values are
expected. This ought to be clear from context, and in any case the worst
that will happen is a BAD INPUT message.

All the programs will operate in continuous loops (REPEAT's), except
for pauses, data inputs and other prompts. Usually, at the end of each
calculation there will be a prompt such as

    S(TOP OR R(EPEAT   ?

to which one replies by hitting the S or R keys. All other keys are ignored.
The effect is self-explanatory. In some programs, similar continuation
prompts may occur in the middle of the program. All these prompts
appear on the text screen.

⟩Modification of programs; library units

The programs are deliberately simple, and users with minimal PASCAL
experience should have no difficulty in modifying them. Modifications
would be needed if for example the user wished to change the choice of
function. To make changes quicker to compile, many of the graphics and
service routines are kept in a separate program unit called FFTSTUFF,
the source text for which is in the file FFTUNIT. Thus all the programs
contain a USES FFTSTUFF statement.

Before compiling any modified programs, you must add the routines
from FFTUNIT to your working copy of APPLE1: SYSTEM.LIBRARY
(make sure you keep a copy on another disc for safety). To do this, first
compile FFTUNIT, then create a new SYSTEM.LIBRARY by copying
the existing library and adding the CODE file made by compiling
FFTUNIT. For full instructions see the *Apple Pascal Operating System
Reference Manual*, Chapter 8. After creating a new LIBRARY you must re-
boot the system.

If possible obtain a listing of the program to be modified before
attempting changes. The only changes that are required by the text are
changes to the PASCAL FUNCTION defining the mathematical function

to be used. This function will be found at or near the beginning of the function and procedure definition block.

Procedures defined in the FFTUNIT are described below. Procedures defined within each program are well commented; the only procedure to be specially mentioned here is SETARRAY, used in FOURIER, DFT and derivatives. This procedure sets up function values in an array F of type CMPLXAR in a format suitable for use by procedure GRAPH (see next section, and §3.4 of main text for an explanation of the format used).

⟩The FFTSTUFF unit

This section is included for completeness; it need not be read if the standard programs are to be run, and will not be required for minor modifications either. There are three groupings of procedures in the unit: graphics, input/output (IO), and the fast Fourier transform procedure (FFT).

FFTSTUFF — graphics
The procedures are

```
GMOVE(C, X, Y)      AXES      SECTOR(XB, YB, XS, YS)
RANGES(XL, YL, XH, YH)      GTEXT(IX, IY)      GRAPH(F, N, R, I, C)
DASH(IX1, 1Y1, IX2, IY2, L0, L1, C)      GDASH(X1, Y1, X2, Y2, C)
```

where the following type conventions are used:

C: screencolour
R and anything starting with X or Y: real
N and anything starting with I or L: integer
F is an ARRAY [0..NMAX, 0..1] OF REAL.

A new type CMPLXAR is created by FFTSTUFF to describe F. Three constants are created: NMAX = 1023, SCWIDTH = 280, and SCHEIGHT = 192. The first controls the size of F, the others are the screenwidth and height. There are 12 real variables associated with FFTSTUFF:

XLOW, YLOW, XHIGH, YHIGH
XBASE, YBASE, XSIDE, YSIDE
XCROSS, YCROSS, XFACTOR, YFACTOR

The heart of the graphic routines is the GMOVE(C,X,Y) procedure.

This acts like the TURTLEGRAPHICS combination

        PENCOLOR(C); MOVETO(X1,Y1)

except that X and Y are in user units. In fact all that GMOVE does is to convert X and Y from user units to screen units as follows:

        X1 = XCROSS + X*XFACTOR
        Y1 = YCROSS + Y*YFACTOR

XCROSS, YCROSS, XFACTOR and YFACTOR may be most conveniently calculated by the AXES procedure. AXES expects XLOW, YLOW, XHIGH, YHIGH to be set to describe the range of user units to be covered by the axes, the origin being assumed within the ranges. AXES also expects XBASE, YBASE, XSIDE and YSIDE to be set to describe the lower left corner (the base) and sides of a rectangle, all in screen units, which is to be the sector of the screen covered by the axes. AXES also draws the coordinate axes.

The ranges may conveniently be set by the RANGES(XL, YL, XH, YH) procedure, and sector data may be set by SECTOR(XB, YB, XS, YS). Normal assignment statements may also be used.

The normal sequence of use would thus be:

        call RANGES and SECTOR
        and/or set the range and sector parameters by
        normal assignment statements;
        call AXES;
        call GMOVE as required.

The procedure GRAPH(F,N,R,I,C) is specifically written to help plot arrays of type CMPLXAR. F is the array, N is an integer, R defines the horizontal range of user units over which to plot, I is the second subscript (convention: 0 real part, 1 imaginary part), and C is the screencolour to be used.

The routine plots the values $F(0,I)$ to $F(N-1,I)$ inclusive; $F(0,I)$ to $F(\frac{1}{2}N, I)$ are plotted at horizontal values from 0 to $\frac{1}{2}R$; $F(\frac{1}{2}N, I)$ to $F(N-1, I)$, and $F(0,I)$ are plotted from $-\frac{1}{2}R$ to 0.

The procedures DASH and GDASH are used to join up screen points using dashed lines. DASH works in screen units, GDASH works in problem units and assumes that XCROSS, YCROSS, XFACTOR and YFACTOR are set. The call for DASH is

        DASH (IX1, IY1, IX2, IY2, L0, L1, C)

The points to be joined are (IX1, IY2), (IX2, IY2) in screen units. L0 is the gap size, L1 the dash size in screen units. If values less than 1 are chosen, default values of 4 and 3 are used instead.

The call for GDASH is

GDASH (X1, Y1, X2, Y2, C)

The default values for L0 and L1 are used.

FFTSTUFF — IO
The procedures are:

PAUSE   QPAUSE   READINT(N, CODE)   READREAL(A, COD E)
VERIFYINT(N)   VERIFYREAL(A)   VERIFYSTRING(A, ASTR)

where type conventions are N, CODE : INTEGER
                                   A   : REAL
                              ASTR : STRING

PAUSE and QPAUSE switch to the text screen when T is pressed, or the graphics screen when G is pressed. Execution resumes only when space or return is pressed. PAUSE prints a prompt before pausing, QPAUSE does not.

READINT and READREAL input integer or real values in the format described earlier. The arguments are called by name (i.e. must be variables, not expressions). CODE is set to indicate the success or failure of the input:

CODE = 10   indicates illegal format, in which case the value of N or A is unchanged.

CODE = 0   indicates a null input; again N or A will be unchanged.

CODE = 1,2 or 3 indicates that input was in integer, fractional, or decimal format, respectively.

VERIFYINT(N) and VERIFYREAL(A) also require variable names as arguments. The current value of the argument is printed in parentheses, and input is awaited. Null input, i.e. an immediate space or return, leaves the value unchanged. Input in illegal format produces an error message and repeats the prompt. Legal input changes the value and exits from the procedure.

VERIFYSTRING(A, ASTR) behaves like VERIFYREAL except that ASTR is printed in place of the current value of A in the prompt. When legal input is typed, the value of A is updated as for VERIFYREAL and ASTR is updated to store the actual string typed. The effect is that if for example the user typed in a value of 1/3 for some variable then the next time the procedure is called for that variable the prompt will be (1/3)?, whereas VERIFYREAL would have used (0.333333)?

FFTSTUFF — fast Fourier transform procedure
This procedure is FFT(F, M, S)

Here F is of type CMPLXAR, namely ARRAY [0..NMAX, 0..1] of REAL, used to hold the real and imaginary parts of the discrete values to be transformed. M is an integer which determines how much of F will be transformed; specifically, F will hold 2^M data values, so the first index will run from 0 to $2^M - 1$. Clearly this value must be less than or equal to NMAX. S is an integer which must be set to $+1$ or -1 (see main text for meaning).

$\rangle$ Program User's Notes

$\rangle$ BBC BASIC Version

$\rangle$BBC Micro and Acorn Electron computers

The purpose of these programs is fully described in the main text. This guide gives details of operation that depend only on the computing and programming aspects.

The programs are

SAMPLE1	FSCOS	FOURIER	FOURIER2	FOURIER3
SAMPLE2	FSSIN	DFT	DFT2	
	FSBOTH			

All programs are written in BBC BASIC, and use the Acornsoft *Graphs and Charts* package for graph plotting. Operating familiarity with the BBC microcomputer is assumed. Programs are supplied on cassette or disc. If possible they should be run on a colour monitor, as this will enhance the impact and educational value of the displays.

Special note for Electron users
The programs have been designed for the BBC Micro but they will also work on the Electron without modification except in the case of programs DFT and DFT2. If you attempt to run these without modification, a 'bad DIM' error message will occur; this is because the Electron does not support MODE 7. The DIM statement in line 60 must be changed to read 'DIM F(1023,1)', and then the programs can only be run for powers of 2 up to 10 rather than 11 as is possible on the BBC Micro.

$\rangle$To run programs from cassette

Make sure that the position of programs on the tape is known, using the

*CAT command. Use LOAD or CHAIN to run the required program in the usual way. If you are using a disc system but are loading from tape, first select the tape filing system by typing *TAPE, then set PAGE = &E00, and then load.

⟩To run programs from disc

Insert the disc in Drive 0, hold down SHIFT, then press BREAK. A menu will be shown allowing any program to be selected. After you have typed your choice, the message 'type SPACE to continue' is displayed; when SPACE is typed, the chosen program is read from the disc, the disc filing system is de-selected (the program automatically issues a *TAPE command), PAGE is reset to &E00, and the program is relocated to this new value of PAGE. The computer is now in exactly the same state as if you had loaded the program from tape. You may either type RUN to run the program normally, or you may LIST or edit the program in the usual way. When you have finished, SHIFT/BREAK will recall the main menu or BREAK will restore the usual disc system.

⟩Input of data

Throughout these programs, data will always be requested one item at a time using either an INPUT statement or *verify mode* (see below).

Incorrect input will usually cause the program to halt; type RUN to go again.

Some programs request input in verify mode, which is useful when a program needs to be run several times with only a few changes to data. After the prompt, a value appears in parentheses. This is the current value for that item of data, and may be left as it is just by pressing the RETURN key. If a new value is required, this may be typed exactly as for a normal INPUT statement, or a valid BASIC expression may be input. The expression is evaluated just as if it were part of the program and its value is assigned to the data item.

Examples of use of verify mode: (user input shown after '?')

Max value for y (2.5)? 2*PI
Coeff of cos(6.PI.t) (.5)? 1/3

Program flow and use of 'escape'
The programs usually operate in continuous loops with prompts such as

Stop or Repeat?

to enable the user to control what happens next Replies to such prompts
should be a single letter.

The escape key halts some of the programs and diverts others to a
prompt such as the above. The latter course is followed whenever a
program requires a non-trivial amount of data, so that verify mode may be
used to repeat the calculation without tedious re-input of data.

⟩Modifying the programs

The main logic of each program is shown in the program outlines in
Appendix 3, which also show the line numbers of the start of each block.
Programs may easily be inspected by using the LIST command with page
mode set (ctrl/N). The most common change that is called for in the main
text is to modify a function; again, see the program outlines for the line
numbers. Graphics displays are done using *Graphs and Charts* Level One;
variations may be made by changes to graphics parameters (consult the
graphics manual).

Some notes in explanation of other commonly used routines follow.

PROCsetF(N,T)
This is used to set sample values into the array F(I,J). $J = 0$ for real parts,
$J = 1$ for imaginary.

Values $I = 0,1,2,...,\frac{1}{2}N - 1$ correspond to values at $t = 0$, T/N, $2T/N$,...,
$(\frac{1}{2}N - 1)T/N$.

Values $I = \frac{1}{2}N$, $\frac{1}{2}N + 1,...,N - 1$ correspond to $t = -\frac{1}{2}T$, $-\frac{1}{2}T + T/N$,...,
$-T/N$. Function values are found by calling FNY(t).

PROCgraph(N,T)
This is used to plot real and imaginary parts of F, using different colours.
The procedure is equally useful for plotting both the original function and
its transform, since the FFT procedure modifies the value of the array
F(I,J). The association of values of I with values of t is the same as for
PROCsetF. When the transform is being plotted, frequency values are

associated with I in exactly the same way, with the frequency range value N/T replacing the time range value T.

The book *Graphs and Charts on the BBC Microcomputer* by R D Harding, is published by Acornsoft Ltd., Betjeman House, Cambridge. A cassette is also available. The author acknowledges with thanks Acornsoft's permission to use these routines in this book.

⟩ Index

Milton Keynes UK
Ingram Content Group UK Ltd.
UKHW031152141024
449569UK00024B/849